W9-BFH-303

THE COMPLETE
WARGAMES
HANDBOOK
REVISED EDITION

Also by James F. Dunnigan

**How to Stop War** (with William Martel)

**How to Make War**

**Dirty Little Secrets** (with Al Nofi)

**Shooting Blanks** (with Al Nofi)

**A Quick and Dirty Guide to War** (with Austin Bay)

**From Shield to Storm** (with Austin Bay)

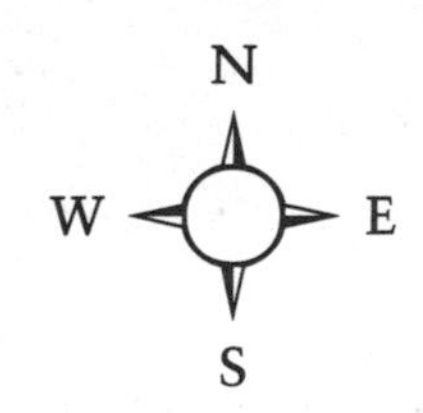

# THE COMPLETE WARGAMES HANDBOOK

REVISED EDITION

*How to Play, Design, and Find Them*

**JAMES F. DUNNIGAN**

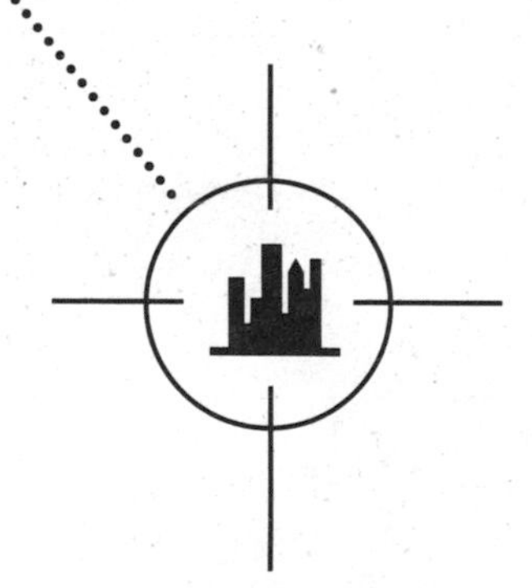

QUILL
WILLIAM MORROW
NEW YORK

It is the policy of William Morrow and Company, Inc., and its imprints and affiliates, recognizing the importance of preserving what has been written, to print the books we publish on acid-free paper, and we exert our best efforts to that end.

Library of Congress Cataloging-in-Publication Data

Dunnigan, James F.
The complete wargames handbook : how to play, design, and find them / by James F. Dunnigan. — Rev. ed.
p. cm.
Includes bibliographical references.
ISBN 0-688-10368-5
1. War games. 2. Computer war games. I. Title.
U310.D86 1992
793.9′2—dc20 92-26553
CIP

Printed in the United States of America

First Revised Quill Edition

1 2 3 4 5 6 7 8 9 10

BOOK DESIGN BY CIRCA 86

To Raymond Macedonia (colonel, U.S. Army, retired), who single-handedly exposed the current generation of generals to wargaming and what it could do. His efforts, aided by a new generation of wargamers and wargame designers in the U.S. military, had a lot to do with the outcome of the 1991 Persian Gulf War, and our prospects for peace into the next century.

# ACKNOWLEDGMENTS

A large number of people helped me create this book and the earlier version. Chief among these groups are all the wargamers I have worked with, for, or at cross purposes to over the years. Wargamers are a sharp lot. Theirs is "the hobby for the overeducated" and they have never shied away from giving me comments, criticism, and advice I needed (even if I didn't particularly want it). Next comes the old SPI crew, some of whom I still work with on a regular basis. It was, among other things, an educational and enlightening experience working in a place one magazine article described as "having the ambience of an unmade bed." The article was from *Cosmopolitan* magazine, so that's how strange SPI was.

For the current edition, I also called upon a number of individuals. Among those I owe thanks to are: Mike Garrombone, Mark Herman, Austin Bay, Ray Macedonia, Brad Anderson, Evan Brooks, Richard Berg, Gordon Walton, Susan Leon, Laurance Rosenzweig, Al Nofi, Dave Menconi, Doug MacCaskill, Brian B. Carlin, David C. Fram, Andrew Webber, James M. Storms, Joshua Willoughby, Lance Jones, Michael L. Malone, Robert B. Kasten, Rolf W. Laun, Robin D. Roberts, and several others whose names I've misplaced. You know who you are.

On the book jacket, the game in the background is *Arabian Nightmare: The Kuwait War,* designed by Austin Bay, published in *Strategy & Tactics* magazine in December 1990.

# CONTENTS

CHAPTERS

# ✷ ✷ INTRODUCTION ✷ ✷

This second edition of *The Complete Wargames Handbook* is back by popular demand. This in itself is encouraging. Twelve years have passed since the publication of the first edition, and a lot has changed in the wargame world since then. The most visible development has been the introduction of personal computers on a large enough scale to allow a substantial shift of commercial-wargaming activity from manual (on paper) to computerized (on PCs) play. A less obvious change has occurred in professional wargaming, as the Defense Department gamers increasingly adopted the techniques pioneered by the "hobby" gamers. Overall, there has been a lot more gaming within the military since 1980, and the story behind that is a fascinating one that is detailed within these pages.

While computer wargames remain a small slice of the overall computer-games market, they have become an overwhelming factor in the wargames market. The growth of computer wargames occurred at the same time (late 1970s, early 1980s) that many gamers were drawn away from history-based wargames by the broader appeal of fantasy and science-fiction subjects. Accelerating the shift to computer games was the appearance of increasingly realistic and jazzy computerized military simulators, "wargames" that put the player in command of a jet fighter, attack helicopter, or warship. These simulators were so attractive that they became more popular than your typical strategic wargame. Despite the appeal of these, though, many potential computer wargamers (largely manual wargamers with a personal computer) were reluctant to buy computer wargames that didn't bear some resemblance to their familiar manual wargames, complete with hexagon grids and so forth.

The net result of all these changes through the 1980s has been a sharp decline in the market for conventional ("paper" or "manual") wargames. Twelve years ago, the average manual wargame sold about 10,000 to 20,000 copies. Some sold a lot more, but the bottom line was that, on average, a wargame sold twice as many copies as the average book on the same subject. Paper wargames were relatively cheap to produce, and during the heydey of manual wargames in the 1970s, nearly 100 titles a year were published. Today, less than half as many titles a year are published, and sales per title are less than half of what they were in the 1970s. Yet wargaming is more popular than ever. It's just that most of the action has moved to computers. The average computer wargame today sells over 20,000 copies per title, and simulator wargames currently sell over 100,000 (and sometimes over 250,000) copies per title. One of the effects of the shift to computer wargames is that, relative to the 1970s, a much larger group of gamers now play a smaller number of games.

Much of this book still deals with manual wargames. This is because even a computer wargame starts out as a manual game. Perhaps a crude manual game, but it's a fact of programming computers that you can't make the computer do something that you haven't first worked out manually.

While I considered devoting this entire book to computer wargames, I quickly realized that readers would not obtain as deep an understanding of wargames as I wished unless I went back to the beginning: to manual wargames. All wargames derive from that earliest and simplest of wargames: chess. Manual wargames are the plateau of realism and complexity chess evolved into before wargames took off into the ever-expanding future of computerized wargames.

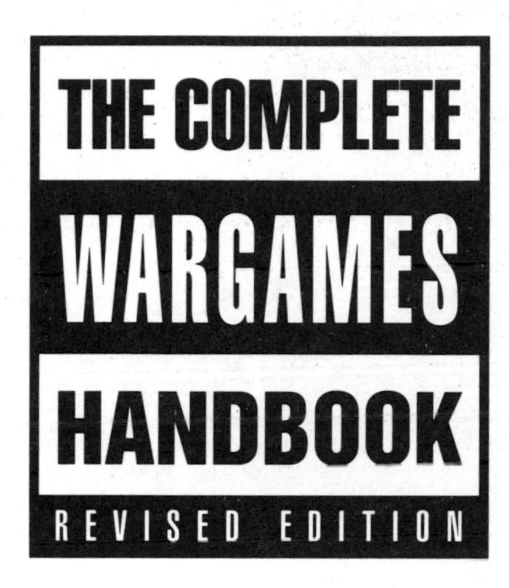
THE COMPLETE
WARGAMES
HANDBOOK
REVISED EDITION

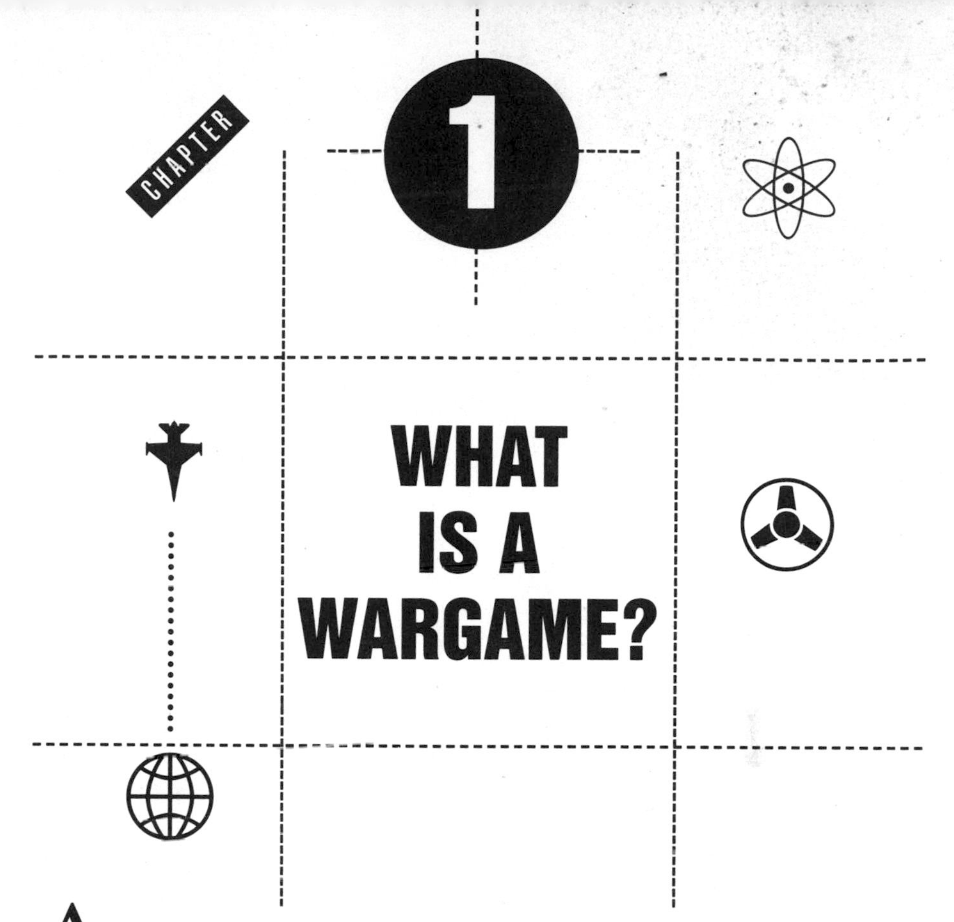


A wargame is an attempt to get a jump on the future by obtaining a better understanding of the past. A wargame is a combination of "game," history, and science. It is a paper time machine. Basically, it's glorified chess. If you've never encountered a wargame before, it's easiest just to think of it as chess with a more complicated playing board and a more complex way of moving your pieces and taking your opponents.

A wargame (manual or computer) usually combines a map, playing pieces representing historical personages or military units, and a set of rules telling you what you can or cannot do with them. The object of any wargame (historical or otherwise) is to enable the player to re-create a specific event and, more important, to be able to explore what might have been if the player decides to do things differently.

To be a wargame, in our sense of the word, the game must be realistic. And in some cases, they are extremely realistic, realistic to the point where some of the wargames are actually used for professional purposes (primarily in the military, but also in business and teaching).

Since the 1980 edition of this book, computer wargames have largely (but not entirely) displaced paper (or "manual") wargames. The personal computer brought a lot of new capabilities to wargames, and the majority of wargamers shifted most of their gaming from paper to computer play once the PC became more available and affordable. In fact, PC ownership by wargamers went from less than one in ten to over two thirds in the early 1990s. The widespread availability of PC-based wargames has created a much larger audience for wargamers. A lot of this has to do with the fact that it's much easier to get into a computerized wargame, as compared to its manual counterpart. Moreover, computers have made possible some types of wargames—namely simulators—that were simply not practical (or possible) as manual games.

Another development since 1980 is the popularity of fantasy role-playing games (FRPGs or RPGs). *Dungeons and Dragons* is the most famous of this genre. These games have their origins in conventional wargames, and the first ones were simply variations on existing games. Many wargamers, particularly the younger ones, dropped the mainline (historical) wargames for FRPGs and then moved on to the computerized versions of FRPGs with the advent of the PC. Through the 1980s, however, it became common to see FRPG players wander back into wargaming, particularly the computerized type. The common thread here was that there was no immense set of game rules to be committed to memory. While computer wargames are a bit more complex to learn than many other types of computer games, they are much easier to learn than manual wargames.

Computer wargames are more difficult to learn than other computer games because wargames are, at heart, simulations of real-life events. A simulation is, by its nature, a potentially very complex device. This is especially true of historical simulations, which must be capable of re-creating the historical event they cover. Re-creating history imposes a heavy burden on the designer, and the player who must cope with the additional detail incorporated to achieve the needed realism. Most computer wargames are also designed to allow the user to play against the computer. This means that the program must have a pretty good artificial intelligence (AI) system. The more recent computer wargames have AI for both sides, and often have the option of letting the computer play both sides, turning the game into a unique form of video entertainment.

In this last respect, computer wargames have come to resemble many of the recent professional wargames. The U.S. Department of Defense (DoD) has been the primary proponent of computerized wargames since the 1950s. Until the 1980s, though, DoD wargames paid little heed to the use of history, as most of their gaming was for battles not yet fought. The older DoD games were of the "black box" variety. You put a lot of formulas and numbers into the computer and got a lot of numbers back. During the 1980s, the military began to study history once more, and now their games

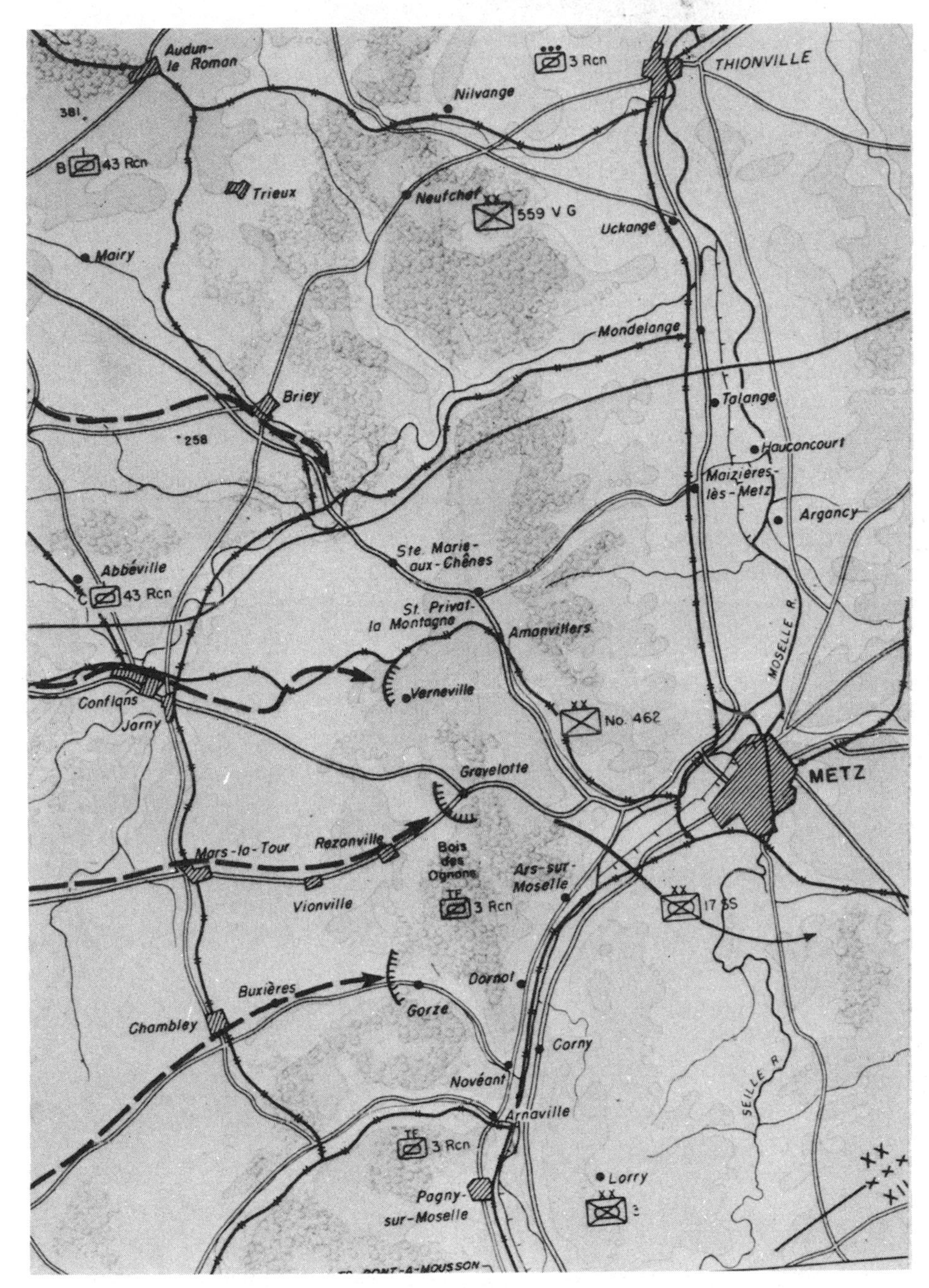


This is the original map from which the wargame map on page 18 was made.

often appear very similar to (although a lot more expensive than) the ones you can buy in a software store. The military also uses a lot of manual wargames, and these too have been heavily influenced by the commercial wargames that became so common from 1960s on. In fact, one reason why the 1980 edition of this book stayed in print so long was because many military schools were using it as a textbook for courses on wargame design.

Wargames come in a wide variety of subject matter, style, and level of complexity. The one you will see in the following pages, *The Drive on Metz,* is a fairly simple one, created expressly for this book. It was designed to do a number of things. First of all, it introduces people who aren't quite sure what a wargame is to what this is all about. Second, it helps me explain to wargamers certain things about how wargames are designed and why they are often done that way. Third, the game re-creates a historical event, General Patton's attempt to capture the French city of Metz in September 1944. What follows is an item-by-item description of what a wargame is.

On page 18 you will find the wargame map, the surface upon which you play the game. The rules on pages 132–143 describe this in more detail.

Most wargames start with a map out of a book (or a map archive or, increasingly, satellite photos) (page 15). Put simply, a hexagonal grid is used to regulate the position and the movement of units (page 17). Each hexagon contains a certain type of terrain. Each type of terrain has a different effect on movement and combat. The game designer analyzes the terrain in each hex and turns the original map into a wargame map. Like chess pieces, the wargame playing pieces represent historical combat units and their capabilities. Wargames involve a lot more numbers than chess, though, so wargame playing pieces have a lot of numerical information printed on them.

| | | | | | | | | | |
|---|---|---|---|---|---|---|---|---|---|
| III 559 1125<br>1-4 | III 559 1126<br>1-4 | III 462 Urfur<br>2-4 | III 462 1010<br>1-4 | III 462 Fhnjkr<br>3-4 | III 3PG 8PG<br>2-8 | III 3PG 29PG<br>2-8 | III 17SS 38SS<br>2-8 | III Metz Garrison<br>1-1 | X 106<br>1-8 |
| III 90 358<br>4-4 | III 90 357<br>4-4 | X 7A CCA<br>7-10 | III 5 2<br>5-4 | X 7A CCB<br>7-10 | X 7A CCR<br>5-10 | III 5 11<br>5-4 | III 5 10<br>5-4 | Game Turn | III 17SS 37SS<br>3-8 |

The pieces represent the military units involved in *The Drive on Metz.* Normally, they would be die-cut from cardboard, colored green for the American units and gray-green (*feldgrau*) for the German units. The

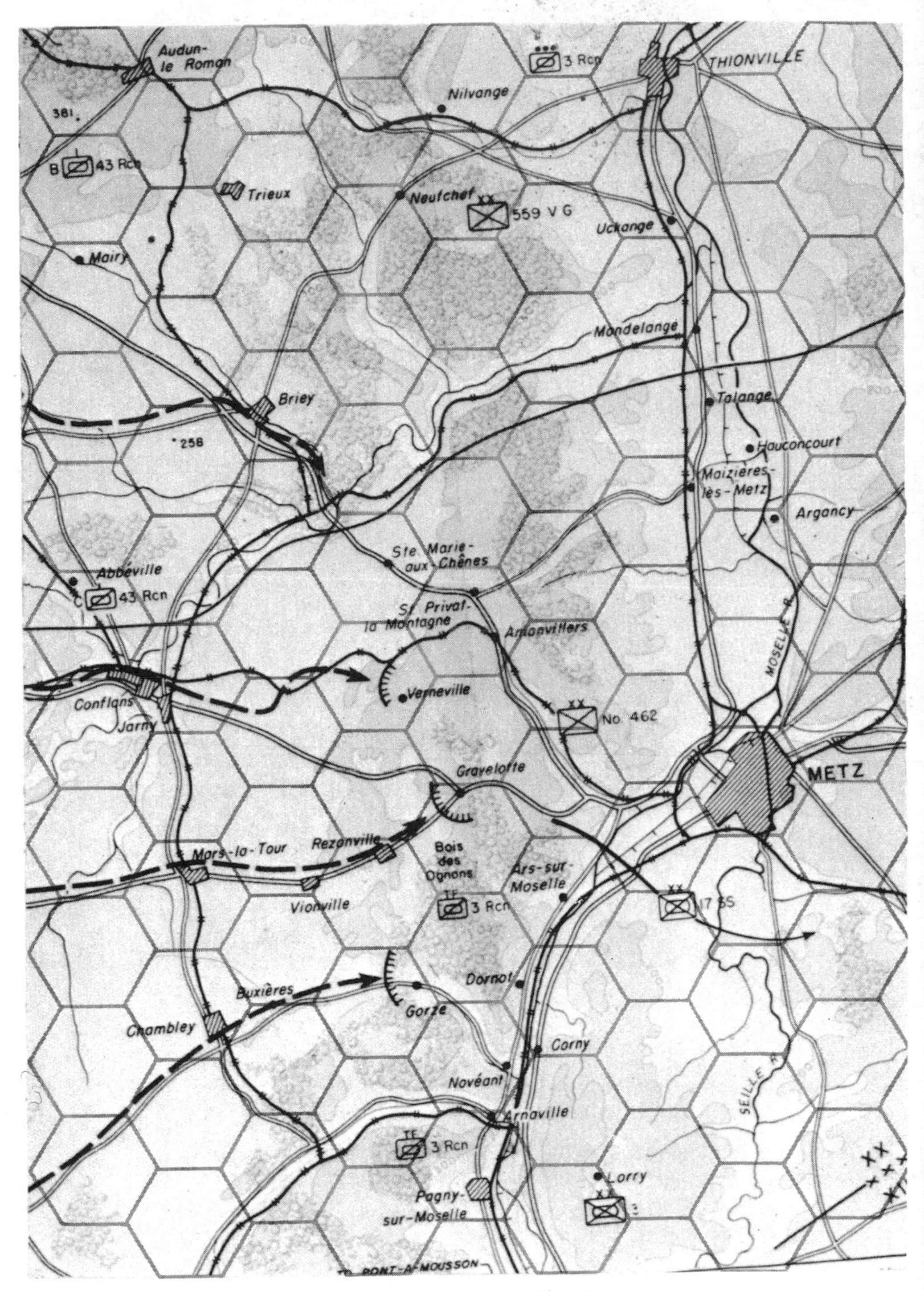


**This is the original map with a hexagon grid overlayed on it.**

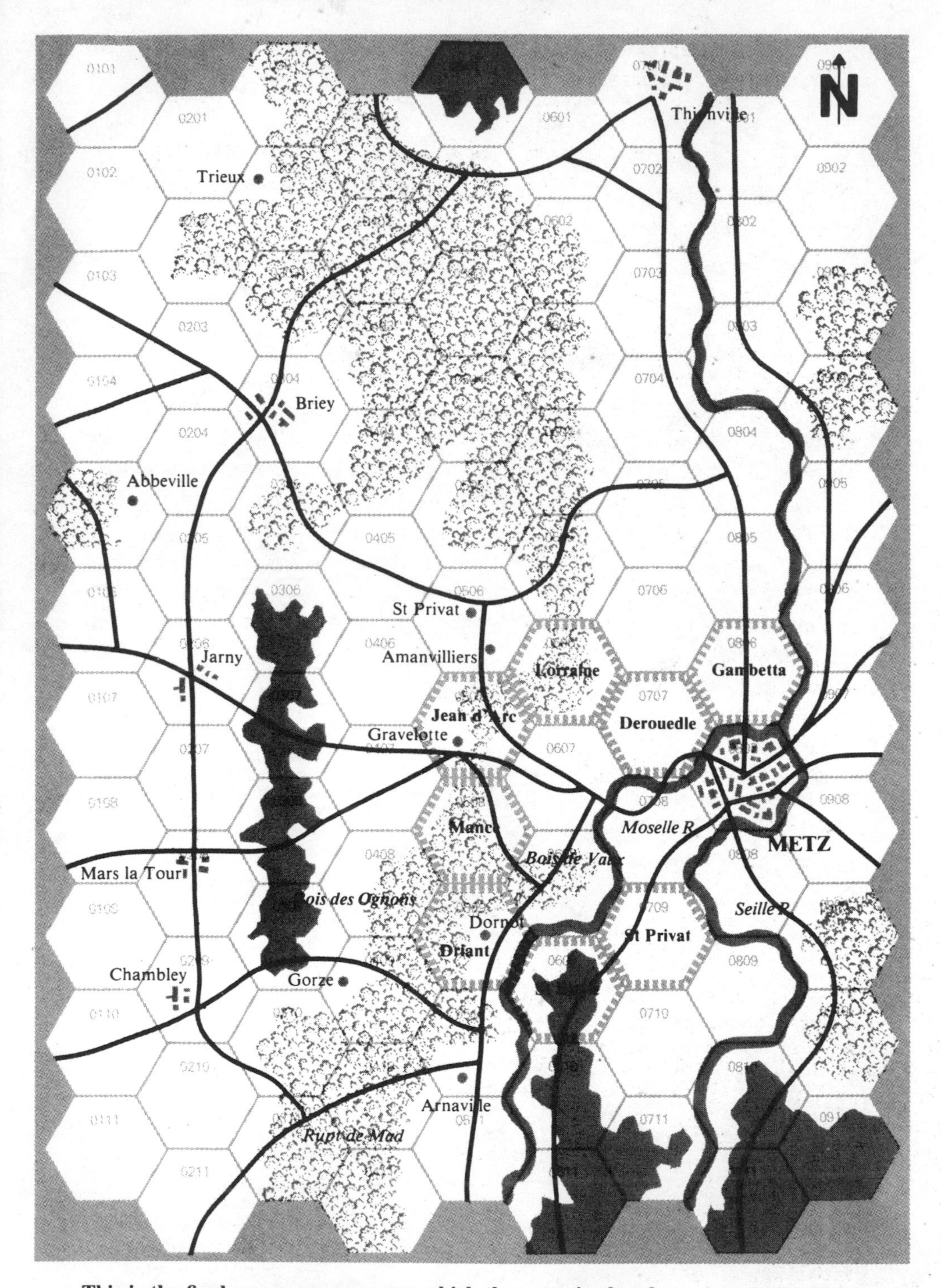


**This is the final wargame map upon which the game is played.**

following illustration from the game rules very succinctly explains what all the symbols on the playing pieces represent.

| | | |
|---|---|---|
| | III | UNIT SIZE [III] SYMBOL [REGIMENT] |
| UNIT DIVISIONAL IDENTIFICATION [5TH] | 5  11 | UNIT REGIMENTAL IDENTIFICATION [11TH REGIMENT] |
| COMBAT STRENGTH [5] | 5–4 | MOVEMENT ALLOWANCE [4] |

There are usually half a dozen or more different types of terrain in a wargame, usually represented by colored hexes on a game map, and each has a different effect on movement and combat. The *Drive on Metz* map shows how terrain differences can also be shown in black-and-white. The Terrain Effects Chart describes the effects of each type of terrain upon movement and combat. As the previous section on the playing pieces described, the number on the bottom right of the playing piece is its movement-point allowance. Each type of terrain costs the unit a different number of movement points to enter. For example, a unit can move much farther on one turn on the roads than it can moving through the forest. In much the same way, the effects of terrain on combat make it easier to defend in woods or towns than in the open. It's actually as simple as it looks. In the past, there have even been chess-game variants that took the same approach.

## TERRAIN EFFECTS CHART

| *Terrain* | *Example Hex number* | *Effect on movement* [MP's to enter] | *Effect on combat* [Leftward column-shifts on CRT] |
|---|---|---|---|
| **Clear** | 0406 | 2 | None |
| **Forest** | 0404 | 4 | 2 |
| **Rough** | 0306 | 3 | 1 |
| **Town** | 0206 | Same as other terrain in hex | 2 |
| **Fortified** | 0507 | Same as other terrain in hex | 3 |
| **Road** | 0405 | 1 | None |
| **River** | 0804 | Must be adjacent at start of movement, uses all MP's to cross | 3 [Only if all attackers are attacking across] |

Finally, we have our Combat Results Table (CRT), which determines the outcomes of battles between opposing units. The CRT also provides for the luck factor so prominent in combat. The phrase "fortunes of war" was invented by soldiers, not poets. Resolving combat is straightforward. When a friendly unit moves next to (adjacent) an enemy unit, it may attack that enemy unit. Success in combat is dependent on two things. The first is the combat-value ratio of the attacker and the defender. The greater the advantage the attacker has in combat value, the greater his chances of being successful.

The second is the element of chance. Not just in warfare, but in most human endeavors, no matter how well we set things up, there's always that strong possibility of something just going wrong. This is why the Combat Results Table is a probability table, in which battle outcomes are determined by the roll of a six-sided die. But luck notwithstanding, the chances of success go up as the ratio of the attacker's force to the defender's increases. In this way, you can make, or at least increase, your "luck."

**COMBAT RESULTS TABLE**

| Die roll | Differential [attacker's strength minus defender's] | | | | | | | |
|---|---|---|---|---|---|---|---|---|
| | −1+ | 0 | +1 | +2, +3 | +4, +5 | +6, +7 | +8, +9 | +10+ |
| 1 | – | DR | DR | DR | DR2 | DR2 | DR2 | DR2 |
| 2 | – | – | DR | DR | DR | DR2 | DR2 | DR2 |
| 3 | AR | – | – | DR | DR | DR | DR2 | DR2 |
| 4 | AR | AR | AR | – | DR | DR | DR | DR2 |
| 5 | AR | AR | AR | AR | – | DR | DR | DR |
| 6 | AR | AR | AR | AR | AR | – | DR | DR |

**–: No result, DR: defender retreat one hex, AR: attacker retreat one hex, DR2: defender retreat two hexes.**

Basically, that's it. Let's go through a few turns of this typical wargame to see how it works in reality. In true wargamer fashion, I played this out by myself, alternately taking each side's situation and making the most of it. I recorded the moves as I went along. While we are doing this, I will also demonstrate how the wargame connects with history. Most of you have read accounts of battles, and they often become quite confusing, even with the assistance of detailed maps. A history-book account of the actions that I played out has been interspersed with the description of what was going on in the game. The "historical" version is in *italics*.

In the following series of illustrations, we will play out two complete turns of *Drive on Metz,* interspersing the game description with equivalent text you would find in a historical account of the battle. Observe these turns carefully, and you will have a sound knowledge of how most wargames work. Comparing the game depiction of the battle with the written one demonstrates why games are considered such powerful (and entertaining) teaching tools. Remember that each hexagon represents four kilometers from side to side, (about 2.5 miles) and each turn represents one day.

## TURN 1: U.S. MOVEMENT PHASE ✱ ✱

*On the seventh of September, Patton's Third Army began its advance into the Metz area. After the breakneck pursuit of the defeated German armies through France during August, the German border was reached. The Allied units were exhausted, and their supplies of fuel and ammunition could not keep up with the troops. As the Germans reached their own border, they were greeted with reinforcements, and fortifications to defend from. Patton's 20th Corps was assigned to seize Metz before the Germans could get organized. The American force consisted of three divisions; from north to south they were the 90th Infantry, the 7th Armored, and the 5th Infantry.*

*The advance was on a broad (45 kilometer) front. In the north, the U.S. 90th Infantry Division's 357th Infantry Regiment advanced on Trieux (31 kilometers northwest of Metz) while the 90th's 358th Infantry Regiment advanced on Briey (23 kilometers northwest of Metz)*

*The units of the 7th Armored Division and the 5th Infantry Division were intermixed. From north to south, these infantry regiments and armored combat commands advanced as follows: Combat Command A advanced on St. Privat (12 kilometers northwest of Metz); the 5th and 10th infantry regiments advanced on Gravelotte (4 kilometers south of St. Privat); Combat Command R advanced 5 kilometers south of Gravelotte; the 11th Infantry and Combat Command B advanced on Arnaville (17 kilometers southwest of Metz).*

*Opposing the two 90th Infantry Division regiments was the German 559th Infantry Division (consisting of two weak regiments, the 1125th and 1126th). Manning the rest of the front (inside the extensive fortifications of Metz) were (from north to south) the German 462nd Infantry Division (the under-strength 1010th Infantry Regiment and the ad hoc officer school, or Fahnenjunker, Regiment) and the weak 8th Panzergrenadier Regiment of the 3rd Panzergrenadier Division. Just to the south of the fortifications ringing Metz was the 29th Panzergrenadier Regiment of the 3rd Panzergrenadier Division. Behind the front-line German units was the 17th SS Division (37th and 38th SS infantry regiments) and the weak regimental-strength Metz garrison.*

On Turn 1, at the beginning of the game, the German units are set up on the map as shown at right. This is how they were actually deployed on September 7, 1944.

The American player decides to make a push in the south. Accordingly, units are moved as follows (from north to south): The 357th Regiment enters the map on hex 0102 and moves into hex 0201. This takes a total of four movement points, which is all this unit has. The 357th Regiment will, during the next turn, proceed to move into the forest, in the direction of Thionville. Because it takes four movement points to enter a forest hex, it will be a slow process, but this will ultimately force the Germans back by threatening to cut off German units in the northern portion of the map. This is a classic battlefield maneuver known as "working around the enemy flanks."

The U.S. 358th Regiment now moves to hex 0104, goes to 0203, and stops in 0304, adjacent to the German 1126th Regiment. This movement took only three movement points, as the unit was moving on roads all the time (and points out how important roads are in warfare). Even the ancient Romans built roads (some of which are still in use) to speed up the movement of their military units.

The American Combat Command A enters the map on hex 0105. This is a road hex, so it takes only one movement point. Combat Command A then goes off the road to hex 0205, 0306, and stops in 0405, expending a total of eight movement points because it was going cross-country most of the time. Combat Command A is now adjacent to the German Unterführer Regiment in hex 0505.

The U.S. 2nd Regiment enters on hex 0106, goes to hex 0206 and stops in 0307. It advances no farther, even though it could, because it wants to allow the U.S. 10th Regiment, which will enter on 0108, to proceed up the road and stop in front of the German 1010th Regiment in hex 0507. This decision shows how the simple game mechanics can re-create the very real problem of traffic congestion in combat maneuvers.

The U.S. 10th Regiment, avoiding the potential traffic jam, enters on 0108, going into 0208, 0308, and stopping in 0407, having expended all four of its movement points by moving on the road at the rate of one point per hex. The U.S. 10th Regiment has now positioned itself in front of a weaker German 1010th Regiment in 0507 in preparation for an attack.

Combat Command R enters on hex 0110, proceeds to 0209, 0309, and stops on 0408, adjacent to the German Fahnenjunker Regiment (in hex 0509), expending seven movement points doing so. The U.S. 11th Regiment, which also enters on the road at hex 0110, proceeds to hex 0209,

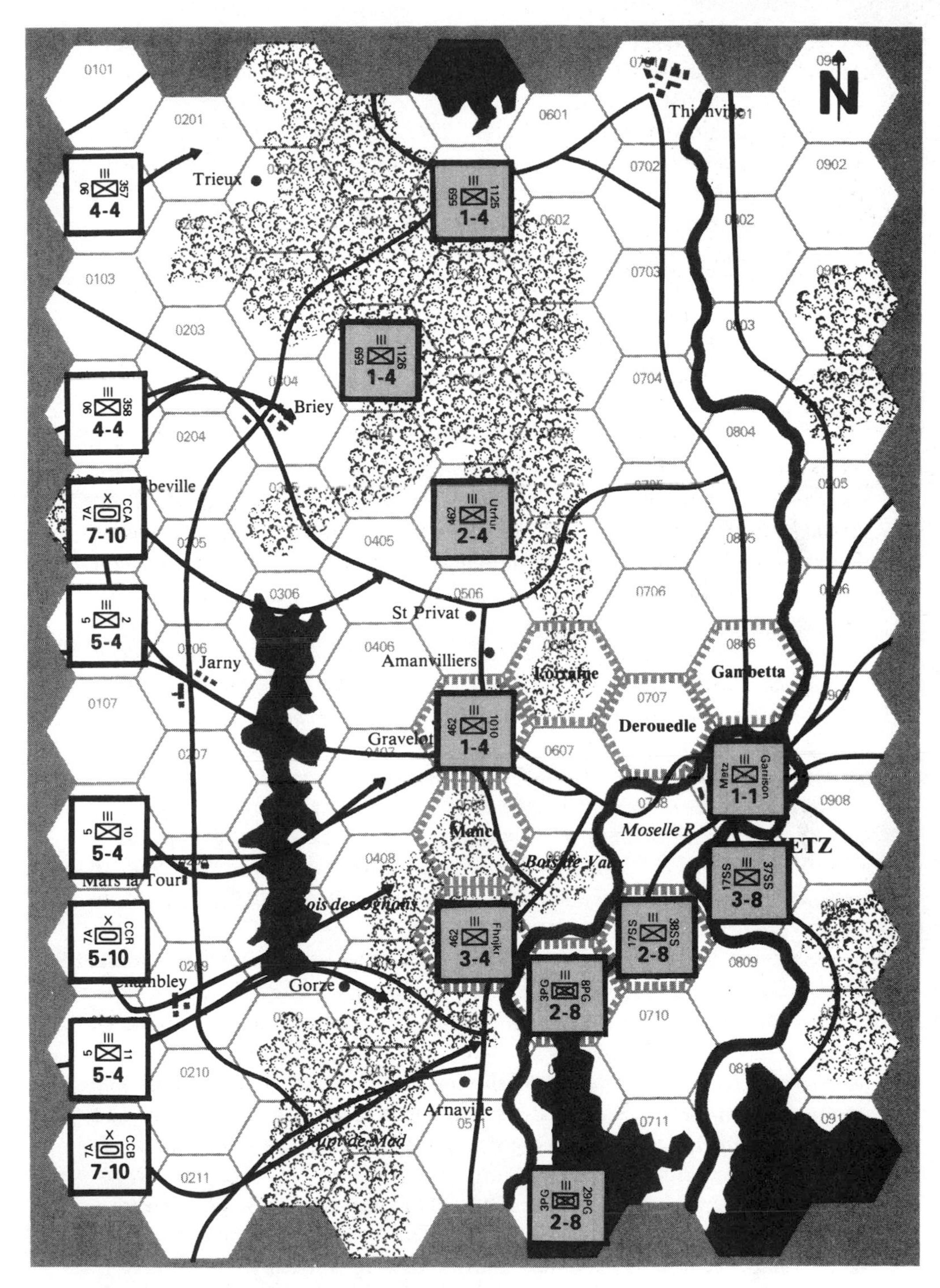


Turn 1: American Player Movement Phase

0309, and stops on 0409, also adjacent to the German Fahnenjunker Regiment.

The U.S. Combat Command B, enters on hex 0111, goes to hex 0211, and then proceeds on the road to hex 0311, 0410, and 0510, expending seven movement points, and ending up adjacent to the German Fahnenjunker and 8th Panzergrenadier regiments. The last three U.S. units are now in a position to attack the German Fahnenjunker Regiment.

All German units that the Americans have moved adjacent to are going to be attacked. Now the actual battle takes place. First we get the story as a conventional history would describe it:

## TURN 1: U.S. COMBAT PHASE ✷✷

*All of the U.S. maneuvers on September 7 led to several battles. In the north, the U.S. 357th Regiment moved toward Thionville, with the main task of protecting the American offensive's north flank. Just south of the 357th, the 358th Regiment ran into the weak German 1126th Regiment, which was deployed in the forest and prepared to resist the American advance. The forest provided ample opportunities for the less numerous Germans to ambush the stronger American units. Faced with this unfavorable situation, it's not surprising that the U.S. 358th was thrown back. The same fate befell the U.S. armored Combat Command A directly to the south. Although the combat command had more firepower than a U.S. infantry regiment, it was even more at a disadvantage attacking Germans in the forest. In this case, the Germans were the Unterführer Regiment, which consisted of students and instructors from a nearby NCO school. These were tough troops, and Combat Command A was unable to push through them. Both German units decided not to advance to the west. They knew the Americans would be back the following day.*

*But these skirmishes in the north were not the main event. Just to the south were the outer works of the Metz fortifications, and this is where the main event occurred on the seventh. It was a mixed success for the Americans.*

*The 2nd and 10th regiments of the U.S. 5th Infantry Division closed in on Fort Jeanne d'Arc, near the town of Gravelotte. Unfortunately, road congestion prevented both regiments from making a combined attack on the German 1010th Regiment holding the fort. Only the U.S. 10th Regiment was able to attack, and the Germans, weak as they were, managed to repulse the American attack.*

*Eight kilometers to the south, the students of a nearby German NCO training school, organized into an understrength Unterführer Regiment,*

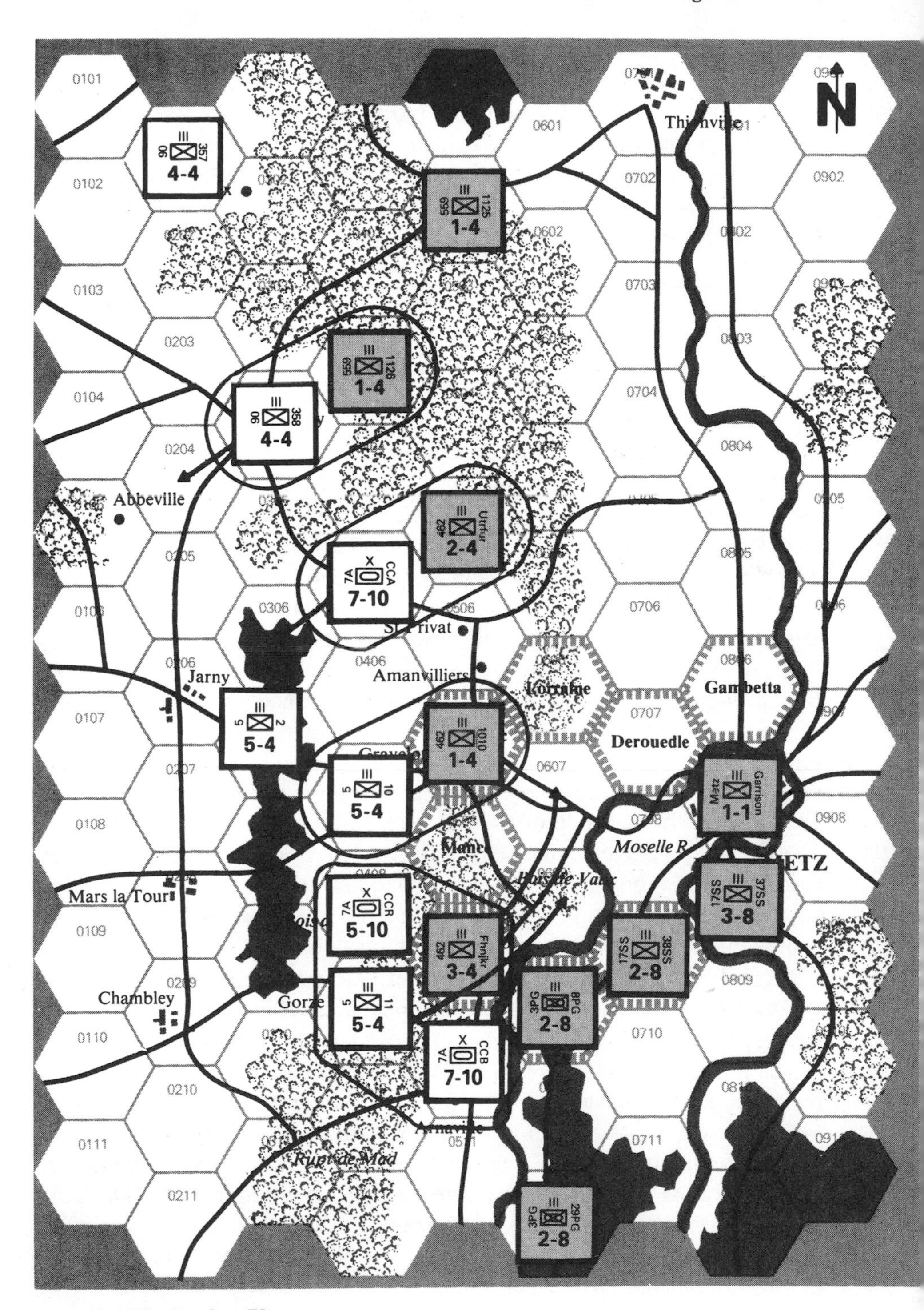


Turn 1: U.S. Combat Phase

*bore the brunt of a concentrated attack by the equivalent of a full U.S. division (the 11th Regiment of the 5th Infantry Division and Combat Commands B and R of the 7th Armored Division.) The Germans were sent back seven kilometers toward Metz, pursued by the U.S. 11th Regiment. The elements of the 7th Armored Division were held back largely because there simply wasn't room to move them forward in the wake of the retreating Germans.*

The U.S. player will now conduct four attacks. Note that whenever possible, attacks are made so that they will put the advancing units in a position to surround adjacent enemy units (assuming the attack succeeds). That way, during a subsequent attack, these surrounded units will be destroyed if they are forced to retreat. Remember this technique, as it's one of the cardinal rules of wargames, and warfare itself.

Starting in the north, the 358th Regiment attacks the German 1126th Regiment. The 1126th Regiment has a combat strength of one. The U.S. 358th has the strength of four. This gives a combat differential (defined as attacker strength minus defender strength) of plus three. Consulting the Terrain Effects Chart, we see that attacks made on units defending in a forest result in a leftward column shift of two columns on the Combat Results Table. This means that the attack will be made on the zero column of the CRT. The U.S. player rolls the die for a five, meaning that he must retreat. The U.S. player retreats into hex 0204. The German player wisely decides not to pursue, because the defensive possibilities are much greater in the forest than outside of it.

The U.S. player now launches Combat Command A against the German Unterführer Regiment in hex 0505. Here the combat differential is plus five, and again, because the defending unit is in the forest, the results of the die roll must be shifted two columns to the left on the CRT. A four is rolled, and the U.S. unit is forced to retreat to hex 0306. Again, the German unit stays put, because it is much safer in the forest.

In the third attack, the U.S. 10th Regiment meets the German 1010th regiment located in hex 0507. This time the combat differential is plus four, but the German unit is in a fortified hex, and the shift is now three columns on the CRT. A two is rolled, which indicates no result, and both units stay where they are. This was a crucial attack for the Americans. If this battle had been won, the U.S. 10th Regiment would have left the German Fahnenjunker Regiment to the south only one hex to retreat into. If the Fahnenjunker Regiment were forced to retreat two hexes (as the outcome of another battle) it would have been destroyed because the game rules do not allow retreats (as a result of combat) across the rivers. As it turned out, the Fahnenjunker Regiment did have to retreat two hexes. But because the German 1010th Regiment held its ground, the Fahnenjunker Regiment was able to get away.

The fourth attack, probably the most critical one, takes place against the German Fahnenjunker Regiment in hex 0509. This unit is being attacked by Combat Command R, the 11th Regiment, and Combat Command B. The total differential is plus 14, which means combat would take place on the plus-10 column of the CRT. However, the defending unit is on a fortified hex and thus the combat results are plotted three columns to the left (on the plus-five column). A one is rolled, and the Fahnenjunker Regiment must retreat two hexes. The only available hex to retreat to is 0607. The U.S. 11th Regiment advances after combat into 0509 and 0608. This completes the U.S. combat phase.

Things have not gone too badly for the German player so far. It could have been a lot worse; if the German 1010th Regiment had been pushed back, the American 10th Regiment would have been occupying hex 0507. This would have blocked the retreat of the Fahnenjunker Regiment (if the American player rolled a one), and this would have busted the German line wide open. It would not have been the end of the game, but the Germans would be hard-pressed to recover.

That situation, by the way, is a classic tactic in wargaming (and war itself): that of setting up one attack with another. But the American player *should* have sent Combat Command A against the German 1010th Regiment in 0507, with two units attacking the 1010th (Combat Command A and the 10th Regiment). Then the 1010th Regiment would almost certainly have had to retreat; and even if a one had not been rolled on the attackers in the south against the Fahnenjunker Regiment (a stroke of luck for the American player), the Germans would still have been knocked out of many of their fortified hexes. This demonstrates yet another aspect of wargaming, that there are many solutions to the historical problem the game presents you with.

*The initial American attack had been surprising in its vigor, but that was what the Germans had come to expect from George Patton, the only American general they considered on a par with the aggressive German commanders. The repulse of the northern hook of the American attack enabled the Germans to thin out their troops up there and build up their reserve farther north. The major battle was obviously going to be around*

*Metz itself; Patton showed his hand by making a forceful, and successful, attack on one of the outlying forts on the seventh. Most of the available German units in the area were therefore concentrated around Metz as the Germans were prepared to make their stand.*

## TURN 1: GERMAN MOVEMENT PHASE ✱✱

The German player decides to take advantage of his good fortune in the north. Both his units up there threw back U.S. attacks, so the Germans can now rearrange the German units to build up a larger reserve. He moves the German 1126th Regiment one hex to the north, to 0402. This allows the 1126th Regiment to cover most of the northern front. The German 1125th Regiment is then moved along the road to 0601, 0702, 0703, and 0704. The 1125th can now be moved either all the way down to Metz or into the vicinity of hex 0505 to relieve the stronger German Unterführer Regiment. It is always best to have your strongest units in reserve when you are on the defensive. For the moment, however, the 1126th Regiment in 0402 and the Unterführer Regiment in 0505 are enough to hold the northern sector.

In the south, the only unit to be moved is the German 37th SS Panzergrenadier Regiment in 0808. This unit will be moved off the map to gain victory points for the Germans. This represents the German player's need to send units to other threatened areas off the map (and outside the action covered by this game). In addition to holding his ground in this battle, sending a certain number of units off the map is one of his primary victory conditions in this game. This is an elegant way of showing how most battles of the 20th century do not occur in isolation, but are interconnected with many other events.

## TURN 1: GERMAN COMBAT PHASE ✱✱

Given the difficulty of obtaining sufficiently high differentials against American units, the German player wisely decides not to attack. His best tactic is to respond effectively with movement only to American movement and combat. This points out the ancient military experience that the defense is the stronger form of warfare. The Germans can get more out of the few troops they have available by not attacking. Of course, the player may ignore this wisdom and, if the U.S. player makes enough dumb moves, German attacks might even make some sense.

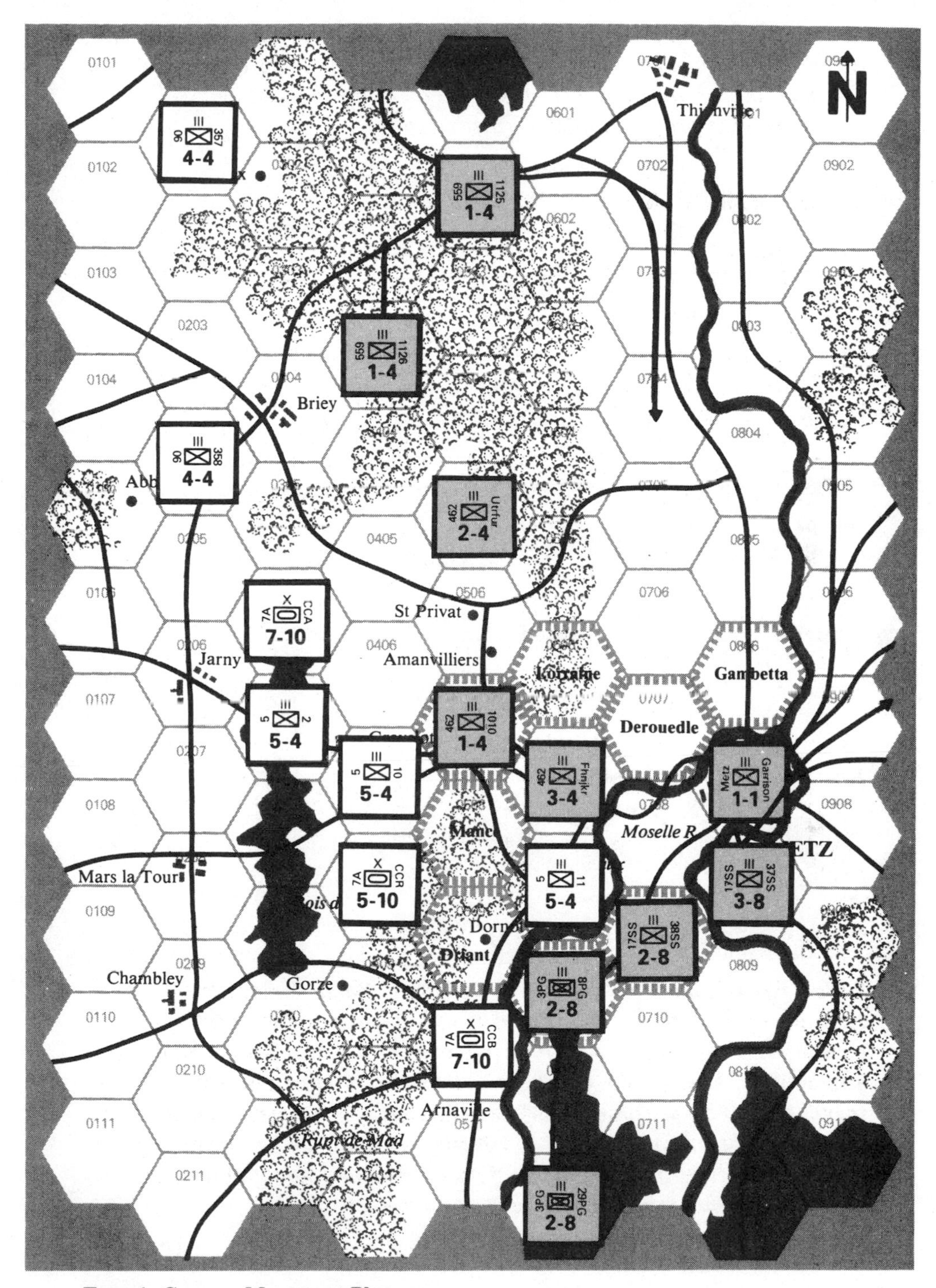


**Turn 1: German Movement Phase**

## TURN 2: U.S. MOVEMENT PHASE ✷✷

*On September 8, U.S. troops were ordered to redouble their efforts to reach the Moselle River and the key city of Metz. Patton noted that the Germans were quite thin north of the city. Suspecting an opportunity to outflank the city from the north, and to ease up the increasing American traffic jam on the outskirts of Metz, Patton moved the 2nd Infantry Regiment of the 5th Infantry Division north to assist the 90th Division's push on Thionville. Their attack was successful, and the German 1125th Infantry Regiment withdrew toward Thionville, closely pursued by the 90th Division. Meanwhile, the 7th Armored Division and the rest of the 5th Infantry Division pushed on through the Metz fortresses toward the city itself. For the second day, the assault on Fort Jeanne d'Arc went on. This time, Combat Command A of the 7th Armored assisted the 10th Regiment of the 5th Infantry Division. Again, the German 1010th Infantry Regiment repulsed the assault.*

*Nine kilometers south, the previous day's successful storming of Fort Driant was carried over to the capture of the Fort St. Blaise complex. The bulk of the 7th Armored Division and a regiment of the 5th Infantry Division fought their way across the Moselle River and drove the defending regiment from the 3rd Panzergrenadier Division eastward. Another regiment from the 3rd Panzergrenadier Division was then detected on the now-exposed flank of Combat Command R, and to its front was the entrenched German 38th SS Regiment. Metz was not going to fall easily.*

*By the end of September 8, General Patton was demanding to know why Combat Command A had failed two days in a row. Meanwhile, intelligence reports indicated that the Germans were moving an armored brigade south toward Metz, and the Metz garrison was ordered out of the city to meet the Americans advancing from the west.*

The U.S. player decides to continue the pressure in the north.

First, the U.S. player goes after the German 1126th Regiment up north. The U.S. 357th Regiment moves from hex 0201 to 0302 and adjacent to the German 1126th Regiment in hex 0402, while the U.S. 358th Regiment moves north from 0204 to 0304 to 0303 to join in the attack. Down in hex 0307, the U.S. 2nd Regiment, which has yet to see combat, is ordered to turn around and march north to assist the drive against Thionville (hex 0701). The 2nd regiment goes from 0307 to 0206 to 0205, 0204, and stops in 0304 because it cannot move far enough in that turn to assist in this turn's attack against the German 1126th Regiment. But next turn, the 1126th Regiment will be in big trouble.

To the south, Combat Command A, unsuccessful in its attack against the Unterführer Regiment in 0505, now turns its attention to the 1010th

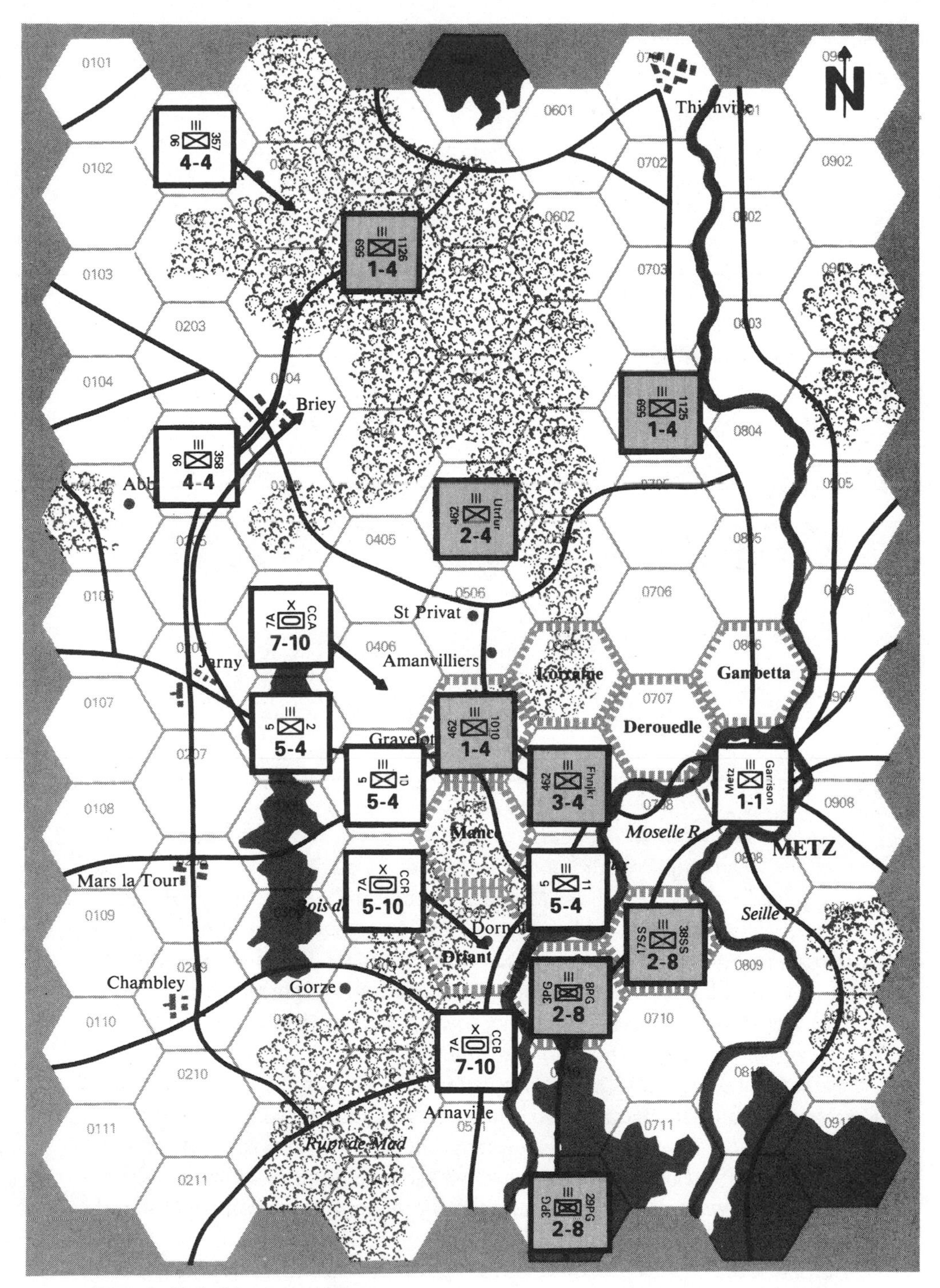


**Turn 2: U.S. Movement Phase**

Regiment in 0507. It moves into hex 0406 in order to assist the U.S. 10th Regiment in hex 0407.

Exploiting the previous turn's attack, Combat Command R in 0408 moves into 0509 in order to attack the German 8th Panzergrenadier regiment. The U.S. 11th Regiment in 0608 and Combat Command B in 0510 will also join in this attack; Combat Command B also holds the southern edge of the U.S. front line. It's usually important in any game to maintain a continuous front, even if you are the attacker. It's rarely a good idea to let yourself get surrounded.

## TURN 2: U.S. COMBAT PHASE ✷✷

The U.S. player now makes three attacks.

In the north, the 357th and 358th regiments attack the German 1126th Regiment. With a high combat differential in the Americans' favor, this whole attack is made on the plus-seven column. But because the German unit is in the forest, the odds shift two columns to the left, putting the attack on the plus-three column. The U.S. player rolls a two, forcing the German player to retreat. The German player retreats to 0502, and the U.S. 358th Regiment advances.

Now there's an important consideration here. The reason the 358th Regiment advanced and not the 357th was because if the 357th had advanced, the 358th, being now directly behind it, could not get adjacent to the German 1126th Regiment for next turn's combat. If the German player had retreated to 0503 instead of 0502, the U.S. player would then have advanced the 357th regiment into the vacated hex, so that the 357th could again move adjacent to the German unit for the next turn of attack. The reason the German player did not retreat southward was because this would have opened up a potential gap to the north that the Americans could have eventually got through, and it also would have made the German unit vulnerable to attack by the U.S. 2nd Regiment in 0304 (which could have been moved into 0403). Thus, you can see how the slightest movement in a wargame can have far-reaching implications, and each move must be carefully considered.

The second U.S. attack is that of the Combat Command A and the 10th Regiment against the German 1010th Regiment in 0507. This attack has a combat differential of plus 11. But again, because the German unit is a fortified hex, the attack is shifted from the plus-ten column to the plus-five column. The American player then rolls the die and has the misfortune of rolling a six, which means that both U.S. units must retreat. Combat Command A retreats to 0306 and the 10th Regiment to 0307. At this point, one would not want to be a member of Combat Command A, given its misfor-

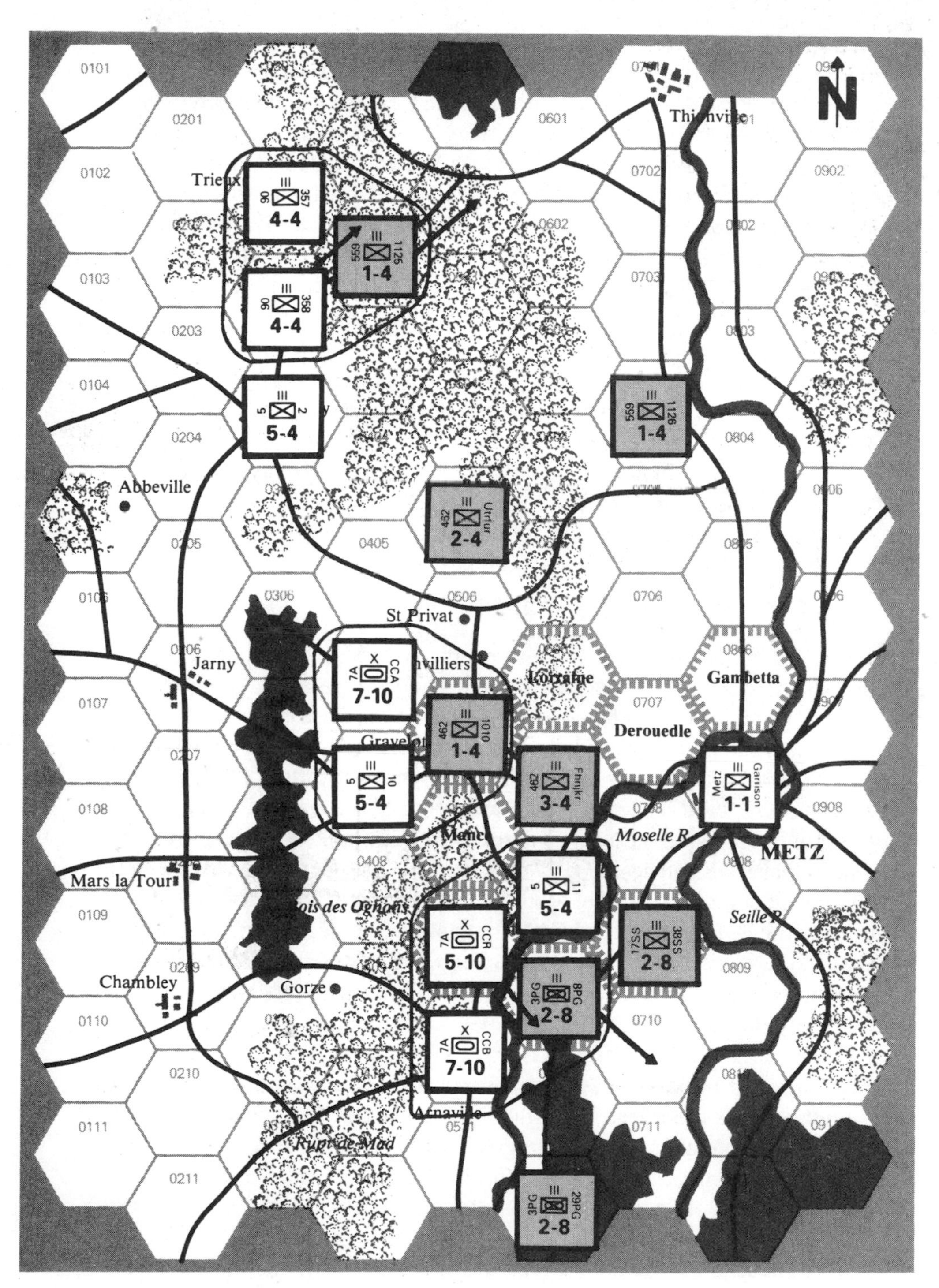


Turn 2: U.S. Combat Phase

tune for being repulsed in two consecutive attacks against two different German units. Equally strange and unfortunate things happen historically, and this is a demonstration of how wargames are able to simulate such events.

The third attack is against the German 8th Panzergrenadier Regiment in hex 0609. The attacking units (U.S. 11th Regiment, Combat Command R, and Combat Command B) have a total differential of plus 15, thus resolving the combat on the plus-10 column. However, the German unit is in a fortified hex, so the battle must be resolved on the plus-five column. A three is rolled. The German here must retreat. The German unit retreats to hex 0710, and Combat Command R advances into the vacated hex at 0609.

The Americans were a bit more successful this time. They successfully put a unit across the Moselle River, and in the north are on the verge of breaking out into the open terrain, thus assuring the seizure of Thionville. This will probably occur by Turn 4. At this point, the German success in the center, particularly in hexes 0505 and 0507, will backfire, as the units holding these two hexes will then be outflanked from the north and quite likely from the south as well, by the continued push of American units there.

## TURN 2: GERMAN MOVEMENT PHASE ✸✸

The German player, aware of the threat the American push in the north and the south is posing, rearranges his units to deal with the situation. The 1125th German Regiment in 0704 moves down the road, through 0804, 0705, and into 0605. The German Unterführer Regiment is now pulled out into reserve, moving into hex 0506 and then to 0605 and 0604. There it must stop, since it has only four movement points, and it took two points to get into 0506 (because it was not entering the hex on a road, it was not able to take advantage of road movement). The Germans get one reinforcement unit this turn, the 106th Panzer Brigade. Before this unit can be moved down into the city of Metz, though, the Metz garrison must be moved out. This unit is moved into hex 0708. The 106th Panzer Brigade then enters the map on hex 0801, goes into 0802, 0803, 0904, 0905, 0906, 0907, and finally into Metz itself in 0807. Meanwhile, on the Moselle, the 29th Panzergrenadier Regiment in 0611 is moved north one hex into 0610. That makes it impossible for the American Combat Command B in 0510 to move across the river. The Americans will now be forced to fight, and fight hard, to make any more progress across the Moselle.

## TURN 2: GERMAN COMBAT PHASE ✸✸

The Germans make no attacks this turn, as again there is no advantage in their doing so.

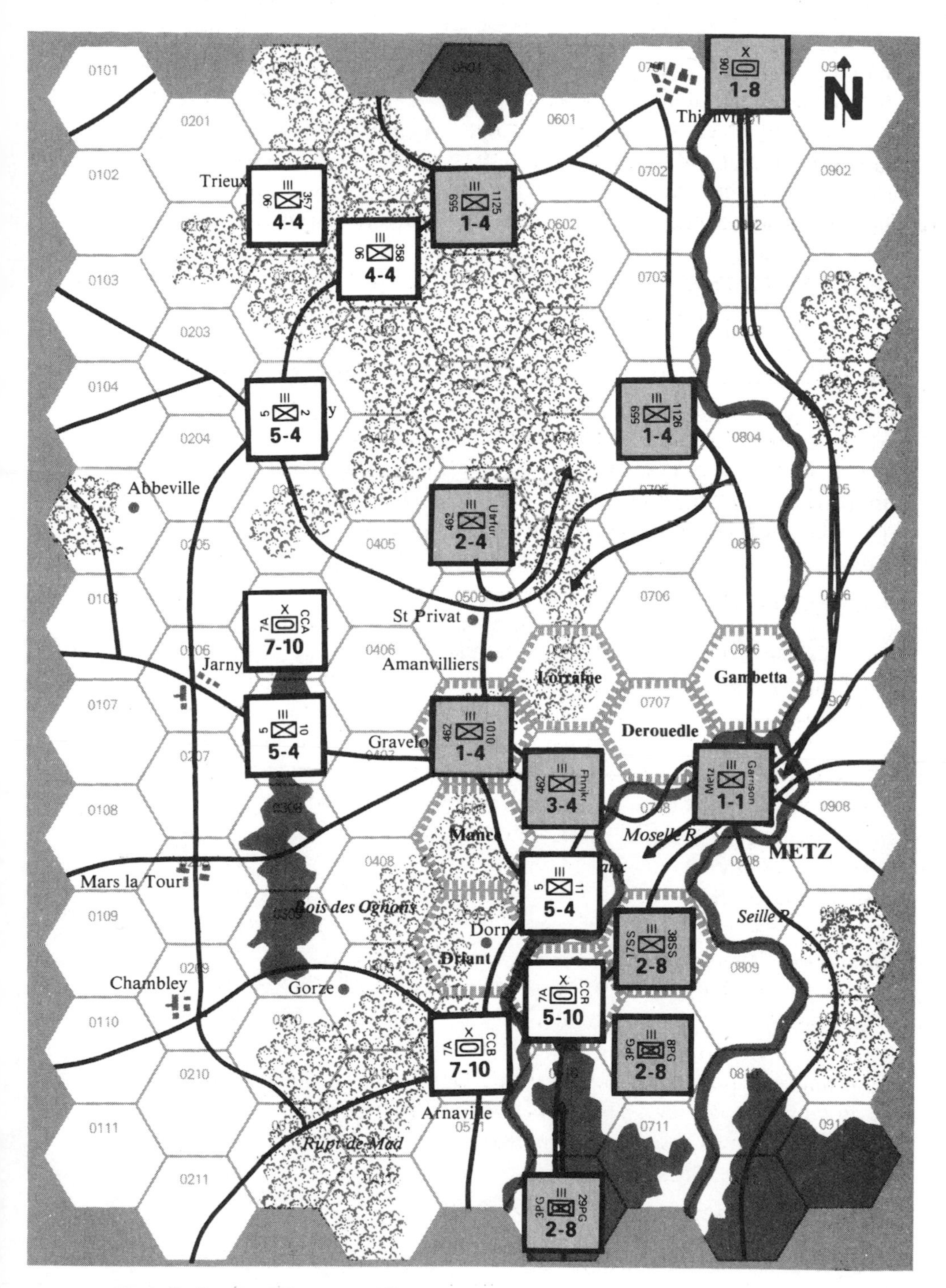


Turn 2: German Movement Phase

Thus concludes two turns of *The Drive on Metz*. The game is only about a third over, and already you can see how it could have developed quite differently if different strategies had been used. You can also see how the game still has considerable potential left in it for a number of things happening. The burden of the attack will continue to be upon the American player. This gives the Americans the advantage of being able to dictate where the battle will be fought. On the other hand, the American player also has the potential for digging a hole and proceeding to fall into it. If you like to take control of things and are willing to accept the risk, being the American player in this game would fit you perfectly.

The German player, on the other hand, merely has to react, and react efficiently. Alas, this position leaves little leeway for mistakes. One or two imprudent moves, and the German player is permanently out of the game. Although it is relatively easy for the German player to make the right moves, there is always the temptation to gamble. Almost all the wargames offer ample opportunity to stick your neck out too far.

If you have not played wargames and want to try one, refer back to this section. The above description captures the essence of most wargames. It is this combination of historical information, dynamic situations, and risk-taking potential that makes wargames so attractive to so many people. What I have just shown you in *The Drive on Metz* is but a tiny sample of what lies beyond.

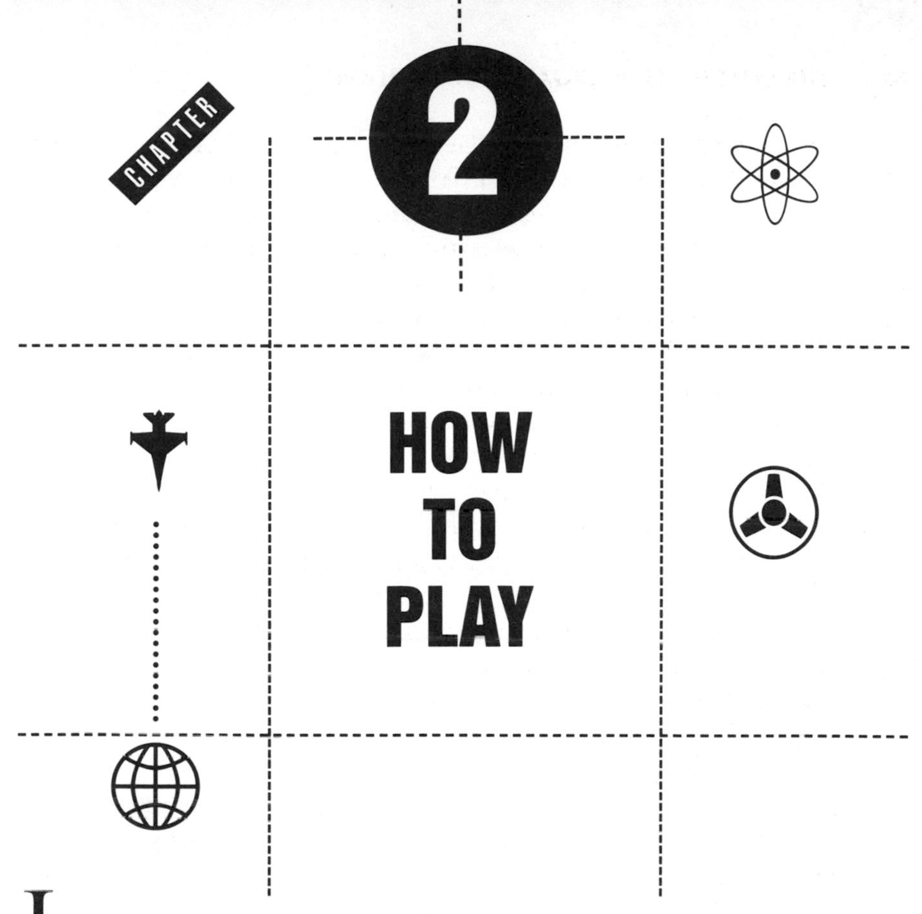


I have found that most wargamers play for a combination of two reasons: to obtain information and to enter a competitive "game experience." The second reason is much the same for any type of gaming or recreational pastime. Obtaining information, however, adds another dimension to a wargamer's "gaming." Aptly called "conflict simulations," most wargames (such as *The Drive on Metz*) are based on historical military conflicts. For this reason, many gamers, being students of history, use the games as learning tools, and also to use their military experience to play the games more effectively.

One assumption that beginning players make is that most people play these games to "win." This is not always the case; many gamers simply play for the experience, to "experience" the information dynamically. But if one is to play to win, one must understand the limitations of the games.

The major limitation of a wargame is that it appears to be a very accurate representation of an actual conflict. The problem is that while the games are somewhat accurate, they are quite inaccurate in certain important ways in order to make them playable. What follows is a detailed dissection of how the games work, for the purpose of being able to play them more effectively, whether to win the game or simply to see what the game is capable of doing.

Note that there is a complete set of rules for the *Drive on Metz* game on pages 132–143 of this book. You may want to consult those game rules as we explore the techniques of playing wargames.

The most common and perhaps the simplest type of wargame to handle is the one using a hexagonal grid across which thick cardboard playing pieces (representing the military units in the battle) are moved. The basic thing a gamer has to grasp is how to move the units most efficiently on the grid, and how to position his units so as to be able to destroy the opposing player's units. Destroying the other side's armed forces is the object of most (but not all) wargames.

In movement there are two key things you must consider. The first is the movement allowance each unit has; that is, the number of hexagons it can move each turn. The next-most-important element in movement is the Terrain Effects Chart, which shows the cost (in movement points) of moving on various types of terrain on the map. In fact, when looking at a new set of rules, these are the first two things you should check. What is the range of movement allowances in the game, and what does the Terrain Effects Chart look like? At this point, you can look at the map and quickly see which areas your units will have an easier time going into and which areas they will more likely get bogged down in. The Terrain Effects Chart also shows the effects of terrain upon combat. Generally speaking, terrain that is more difficult to move in is also more difficult to attack in. In most games, this fact is represented by a unit defending in such terrain being able to increase its defense strength by a factor of two or three or more. For example, a unit with a combat strength of four defending in good defensive terrain would have an effective defense strength of, say, eight or even twelve, depending on the severity of the terrain.

At this point, without even getting into the rules too deeply, you have an idea of how the various units (of both sides) are able to move about the map and their chances of success in combat in the various kinds of terrain. Now the next thing you need to look at is the Combat Results Table (CRT), which translates the roll of a six-sided die into victory or defeat for each attack. Almost all games have these. Some games have many of them, but it's simply more of the same thing. Most Combat Results Tables are based on the idea that the more combat power you have, the better your chance of success. There are usually two broad types: bloody and bloodless. The bloody CRT generally involves a lot of results that have units being eliminated from the game. A bloodless CRT is one in which the losing units are usually moved back (''retreated'').

Now you can start reading the rules of the game. The rules will make a lot more sense because everything else in the rules basically revolves around nuances of moving and fighting with your units.

When a game is designed, generally the first things that go into it are the map, the values on the playing pieces (combat strength and movement allowance), the Terrain Effects Chart, and the Combat Results Table. As the game is developed, various bits of what I call "chrome" are added. You must now search through the rules for these other elements, since they will affect, sometimes dramatically, the use of the basic game elements (terrain effects, CRT, map, counters). You don't really have to study the rules in scrupulous detail. Once you have an understanding of what is supposed to be going on in the game, you should immediately set the pieces up and have a go at it. And this can or should happen 10 or 15 minutes (if not sooner) after you start studying the basic components, or later, depending on the size and complexity of the game.

Most games come with two or more "set-ups." One of these is usually the historical set-up and gives the position of the units of both sides at the beginning of the battle so that you can attempt to re-create the battle yourself. It is generally a good idea to set up the basic historical situation. This will do two things for you. First, it will give you a bit of historical information, which is one of the reasons why you got the game in the first place. Second, it will give you an opportunity to look at some interesting situations that will help you learn the basics of wargaming tactics.

Once you have this historical set-up, you should take one section of the map containing no more than, say, three or four hexes on a side, and proceed to practice the movement and combat as outlined in the rules. You should take only a small section of the map, because you are liable to make mistakes. You will undoubtedly constantly want to restart the game as you find you have made mistakes, so it is better to deal with only a small number of units.

Now it's a good idea to consider the possibility of adding a few enhancements to your game components. Take, for example, the set-up locations of your units. Generally, it's a hex. In a few games, the hex is marked with a symbol showing the unit that's supposed to go there, or the unit itself has the hex number printed on it (if numbered hexes are used) for that unit. If this is not the case, you might consider taking a very fine-point pen (so you can write small) and write either on the front or the back of the playing pieces the hexagon they are normally set up on, or perhaps even marking the map. This is so that when you go through a turn or two, either in testing the game or actually playing it, you can easily go back to "Go" and start again if you find that you would like to have done something differently.

The biggest cause of mistakes is overlooking some minor, but critical, modifications in the rules that invalidate two or three turns of your playing. Typical examples are games that have a rule whereby one side is limited in its movement for the first two or three turns of the game or where on certain turns of the game, certain units on one side or the other have a bonus, or where certain terrain features change in their effectiveness. These are all "chrome" elements that, quite frankly, often don't have that dramatic an effect on the game itself, but that, if ignored, allow the players to be creative and play "their" way, rather than the way in which the designer intended. This is something that is easier to discover if you take the game in small pieces, a chunk at a time, as it were. Keep in mind that a full-size (22-inch-by-33-inch) map sheet can contain as many as 2,000 hexagons. In most games using a map of this size, the number of hexes that will be most actively played over will amount to as many as 300 or 400. For this reason, it makes good sense to choose a small area with a small number of pieces to test and develop your knowledge of the game before committing yourself to moving all those playing pieces over all those hexagons.

New gamers should be made aware of the fact (before they actually learn it for themselves) that the game does not have to be played the way the designer intended it in order to be played well. While most publishers make a fetish of at least attempting to come out with well-written and complete rules for their games, these efforts often fall short (often far short) of the goal. Yet many games with truly terrible rules continue to be popular and widely played. The reason for this is simple. The average gamer is a rather intelligent person who has immersed himself in the basic mechanics of wargames (and we just reviewed most of them) and has a keen interest in and knowledge of historical events. Such a person can take a game with mangled rules but good intentions and turn it into what the designer intended. The players can and do change things themselves in games and get away with it. Most games are not even played strictly according to the rules. I have seen this happen many, many times. Yet the games still play. This is less likely to happen with a computer wargame, but even here, programmers are often astounded at how gamers managed to get around the way the program was supposed to work.

Depending on which way you bend the rules, the game will play better or worse for it. A really good game is usually dependent upon one of two factors: either very high quality in design and development (good rules to begin with that are accurately and easily presented) or simply good players who are interested and excited enough about a particular game to make it work and work well.

Most games are purchased and played on the basis of their subject matter (more than any other characteristic). If you have an abiding interest in the subject of a particular game, you will, without much prompting, make the game do something interesting.

## HOW TO WIN ✷✷

Now that you've got the basic stumbling blocks out of the way, we can get down to some real, nitty-gritty how-to-win advice. While the game's victory conditions drive the game, the nuances of the game rules are what will get you where you want to go.

The first thing you'll want to do is "crack the system." Every game has certain idiosyncrasies that determine the "flow" of play. Such an idiosyncrasy would be, for example, one side having a larger number of weaker units than the other. Another example would be one side starting the battle with its units spread over a far wider area than the other. You should make a note of things like that. Next, go back and examine the Combat Results Table again. Most CRTs employ the odds system; that is, you total up the attacking player's strength points (numerical value) and compare them to the defending player's. You divide the defender's strength points into the attacker's and get a ratio that is usually rounded off so that 1.2 to 1 would be 1 to 1 or 3.2 to 1 would be 3 to 1, etc. Another common type of CRT is the differential system, where you simply subtract the defender's strength points from the attacker's, in which case you get a superiority (+1, +2, +3, etc.).

In most CRTs of both types (in fact in practically all) there is a "break point," a particular column on the table above which the attacker is favored in battle. In odds-based CRTs, this is generally the 3-to-1 column. In other words, if you launch an attack where you have three times as much combat strength as a defending unit, your chances of success are good. As the odds go up (4 to 1, 5 to 1, etc.), they get better. You must examine the CRT of the particular game that you are playing and note where the advantage begins. Thus, when you're playing around with a small section of the game map and units, you can see the types of attacks you are most likely to get. Again, this involves all of the key elements of the game. In order to examine the CRT, you must take into account the Terrain Effects Chart, the general types of terrain you will be encountering in playing the game, and the combat strength and movement ability of the units.

One of the things that makes most wargames different from more abstract games, such as Monopoly, is that they go out of their way to pay attention to what actually happened historically. This means that since most battles were the result of one side or the other underestimating its opponent's strength, there was often a disparity in the strengths and abilities of the two armies. In other words, one side usually has a tough row to hoe. Its armed forces are outnumbered. They're outclassed. They're going to have to work very hard in order to win the game. Sometimes the victory conditions in the game reflect this. The potentially losing side may be granted a ''game victory'' if it does not lose as badly as it did historically. But devices such as this still do not change the fact that the inferior side is going to be constantly reacting to the activity of the superior side, and ultimately losing most of the time.

Let's review the steps that we have taken so far to turn this mysterious jumble of numbers, hexagonal shapes, colors, and charts into something recognizable.

**One:** We have examined the playing pieces, Terrain Effects Chart, map, Combat Results Table, and victory conditions for patterns of information.

**Two:** After examining these, we have gone through the rules to find whatever modifiers there are to these patterns.

**Three:** We have set up and examined the game situation, usually using the historical set-up, and have studied this, if only for the historical information it reveals.

**Four:** We have taken a small section of this historical set-up and begun moving the units around, having combats and generally getting a feel for the game.

**Five:** We have taken the Combat Results Table and analyzed it for the break point where it favors the attacker over the defender.

You now have the tools to begin cracking just about any game. Some games will involve a lot more work than others, but these are the basic techniques. Now to use them.

## HOW TO WIN WITH THE LOSING SIDE ✱✱

Assume you are the superior player in a game; that is, you are the one with most things in your favor (number of units, strength of units, positions, etc.). You have two options: Have a lot of fun being aggressive and outrageously innovative and take a chance on losing the game, or be cautious and have a pretty-much assured, albeit dull, victory. If you are the inferior player, you have only one choice: Dig in, use your head, and have an interesting game. You are at a disadvantage: You can't afford to make mistakes, and you will be faced with one challenge after another. This is why many players prefer the defending side, since with the superior side it is too easy to get lazy, become adventurous, and feel foolish for losing a game that should have been an assured victory. On the other hand, simply settling down into a dull, predictable pattern in which you grind the other fellow into defeat is no fun either. The inferior side in any game is generally more exciting. Besides that, you already have an excuse for losing, and you feel a lot better if you happen to win.

Once you have comprehended the basic principles of any game, it is time to decide which kind of gamer you are. Put somewhat crudely, you are going to play either to win or to learn history. Always keep in mind that what constitutes victory in a historical situation is somewhat more vague than what is presented so clearly in the game.

If you're going to play the game primarily to experience history, I refer you to Chapter 3, "Why Play the Games," which explores in some detail what bits of history to look for in a game. But the game is still a game, and you will find your "historical experience" considerably enhanced by the use of many of the techniques found in this section. For this reason, the first thing I have discussed is how to basically analyze a game. Much of the rest of this section consists of further analysis and tricks of the gaming trade.

Once you have learned the game, your next step is to discover the "play-cycle pattern" of the game. Each game, because of its unique combination of elements, has a certain dynamic about its play. This dynamic manifests itself in a pattern of how most campaigns in that game will go. For example, take a game in which one side (like the Americans in *Battle of the Bulge*) is considerably weaker than its attacking opponent (in this case, the Germans). As the game progresses, however, the weaker defending side continually receives far more reinforcements than the attacking side. This usually means that the attacking side must win early if it is to win at all, before the defender's inexorably growing strength makes a stalemate quite likely and defeat of the attacker quite possible.

In some games, the play-cycle pattern is quite obvious. In many games, though, it is a bit more complex. For instance, one side may have an advantage in one section of the game map while being at a disadvantage in another. Even more likely is a situation in which one side has a potential advantage that can be realized only by making the correct preliminary maneuvers before beginning the battle. Even more subtle are those games in which one side's advantage is not immediately obvious, but is buried under layers of special rules and conditions of the game. Again, this type of advantage can be uncovered and put to use only after careful analysis of the game. Remember that the play-cycle pattern is an important bit of historical insight you don't normally get out of a history book. It's another example of the kind of historical information that can be obtained only from a wargame.

The easiest and most straightforward way to discover the play-cycle pattern of a game is to simply sit down and play the game solitaire a few times. If not the whole game, then at least the same sequence of turns over and over again. The opening turns of a game are usually quite dramatic, and these can be profitably played two or three times.

It's very important to analyze a game through solitaire play. For one thing, you can concentrate. It's just you and the game. There are also psychological elements. There's no pressure on you to win, although, let's face it, you can still lose playing against yourself. But if nobody's there to see it when you screw up, you learn from your mistakes without suffering the ridicule of others (unless you enjoy dwelling on your defeats in public). You also don't waste another player's time, as your inept moves provide no challenge at all for your opponent.

The play-cycle pattern emerges from the total interaction of all the things going on in the game, and there are, even in the simplest games, a tremendous number of things going on at once. As a gamer, you must be able to handle this multitude of elements in your head at the same time. It's much like playing a musical instrument. Musicians don't consciously command their eyes, ears, fingers, lips, whatever, to do all of the coordinated things that must be done to produce a musical sound. They simply play. Wargamers bring to the games a similar aptitude for examining the various detailed actions that go into the play of the game, simultaneously, and then sorting out the really useful information.

People who cannot play the games (as well as many who can but have not bothered to try them yet) are totally overwhelmed when first confronted with a wargame. Many people overcome this initial state of shock by applying their historical knowledge, one of the key reasons why many people get into the games in the first place. This historical approach often leads to some interesting game results. The actions of the historical commanders were not necessarily the most efficient, but at least repeating history gets you into playing the game.

## OVERCOMING MATH ANXIETY ✷✷

The most efficient way to approach any game is by analysis. This does not involve a lot of mathematics or number-crunching. It's simply a question of knowing what to look for. The things you look for are "adjusted ratios."

Since any wargame is based on numerically expressed elements (e.g., attack and defense strengths, movement allowances, etc.) to thoroughly analyze the game, you need to perform three or four basic calculations.

The first one is to determine the total combat strength of each side. This will not be as easy as it sounds, because in many games there are reinforcing units, so you must adjust the number of strength points to the number of game turns that each strength point is in the game. The easiest way to do this is simply to multiply each strength point by the number of turns it will be in the game as we've done here for one of the many Battle of the Bulge games. (You assume the unit will not get destroyed, but we'll get into that later.)

**ADJUSTED RATIOS FOR BULGE GAME (SPI)**

| | *German* | | | *American* | | |
|---|---|---|---|---|---|---|
| Game Turn | # units | Raw Str | Str x Movement | # units | Raw Str | Str x Movement |
| 1. | 20 | 57 | 415 | 12 | 24 | 155 |
| 2. | 20 | 57 | 415 | 15 | 37 | 265 |
| 3. | 20 | 57 | 415 | 18 | 51 | 335 |
| 4. | 24 | 74 | 585 | 19 | 55 | 355 |
| 5. | 25 | 75 | 590 | 21 | 65 | 455 |
| 6. | 25 | 75 | 590 | 25 | 81 | 535 |
| 7. | 26 | 77 | 600 | 26 | 87 | 595 |
| 8. | 29 | 86 | 680 | 28 | 92 | 620 |
| 9. | 29 | 86 | 680 | 28 | 92 | 620 |
| 10. | 31 | 90 | 700 | 28 | 92 | 620 |
| 11. | 31 | 90 | 700 | 31 | 103 | 695 |
| 12. | 31 | 90 | 700 | 31 | 103 | 695 |
| 13. | 32 | 92 | 710 | 31 | 103 | 695 |
| 14. | 32 | 92 | 710 | 33 | 110 | 750 |
| 15. | 32 | 92 | 710 | 33 | 110 | 750 |
| 16. | 32 | 92 | 710 | 33 | 110 | 750 |
| 17. | 32 | 92 | 710 | 34 | 114 | 770 |
| 18. | 33 | 94 | 720 | 34 | 114 | 770 |

**Str = Combat strength. Chart shows cumulative number of units in game each turn as well as total combat strength with and without adjustment for movement ability.**

This, in most cases, will give you totals for the two sides that are not too dissimilar. If the totals are quite different, this is probably mitigated by other elements such as terrain, or perhaps an enormous discrepancy early in the game (meaning that at that point many of the defender's units would be almost automatically destroyed), or other elements having to do with the set-up of the game. As the Bulge game proceeds, the Germans are forced to attack in order to achieve their victory conditions. As always, the defense is easier than the offense, and as the Americans get much stronger as the game goes on, eventually it is the Germans who are at a large disadvantage.

If you really want to get into it, you can extend this analysis to a turn-by-turn basis. Indeed, many of the techniques I am describing here are the ones I use in designing games. When I play them, I must confess, I tend to play them as a historian. Like many gamers, I use the games to gain an insight into a situation, not to improve my won/lost record.

In addition to being modified by the number of turns a unit is present in the game, combat strength is altered by other elements (primarily the Terrain Effects Chart, i.e., the type of terrain the defender would be in). At this point, your analysis will not be so cut-and-dried, since the player who is defending may choose to defend any number of places. It is up to you to examine the defender's situation and, in light of the victory conditions and the lay of the land, determine what the most likely defending positions would be. The average of their additional advantage to the defense would then be added to the amount of strength you have already calculated for the defender.

Not all games are that simple. In fact, the games that are most in need of this type of analysis often have other complications such as weapons that fire over a number of hexes or games in which air power and naval power are also included. Other elements that complicate your life are games with command-control rules (the player is not in complete control of all of his units all of the time) or rules for such things as disease (units will more or less randomly disappear as a result of disease or other types of noncombat attrition). It is things such as this that keep the games from becoming stereotyped. As you can see, a player can devote considerable time to the study and analysis of a game as a game. However, no matter how much you analyze the game from a technical point of view, you are still analyzing the historical situation, since the designer put these pieces together with the intention of achieving a desired historical result with what seemed the appropriate historical effects. Any good game, by definition, will not only have a historical outcome but also will use historical effects to achieve that result.

Analyzing the value of a unit when the unit's capabilities are more than simply a number printed on it is the same type of problem that has faced military commanders throughout history. To give you a current example, consider the problem of determining whether American, German, or Russian tanks are superior to one another. None of the tanks have in common any one factor that will guarantee accurate prediction of the tank's relative strength over another tank. There are several major elements: mobility, armor protection, fire control, effectiveness of the main gun, size, tactical deployment, mechanical reliability, etc. The contemporary military, faced with this problem, has no better techniques to solve its problem than you have to solve yours. Note that several wargames on the 1991 Iraq war were designed and published before the coalition counteroffensive commenced on January 17, 1991. I was involved with one of these games (*Arabian Nightmares*), and that game did accurately resolve the problem of whose tanks were better and to what extent.

To sort out all the possibilities in a game on a battle not yet fought (as in *Arabian Nightmares*, which was completed by early October 1990), you simply play the prototype game to evaluate the various elements in it. Before you start, you take the latest historical information (that is, anything that has happened up to the present) and assume that the troops won't perform much differently tomorrow. Of course, this gets shakier if you are fighting a battle 10 or more years in the future. But when we were creating *Arabian Nightmares*, we knew where the coalition and Iraqi forces were in terms of track record and capabilities.

This still leaves the problems of how new technologies have changed the way these future battles will be fought. A game set in World War I (where radio communications were scarce and air power weak) would have to deal with the situation differently than in World War II (where both items were much improved). The 1990s brought us satellite communications, precision bombing, and invulnerable U.S. tanks. What we had to do was figure out what effect the new technologies had on the standard wargame elements (combat power and mobility, for starters).

To sort all this out, you simply make some top-of-the-head estimates, put together the prototype game, and then play it to determine which elements are critical and to what degree. While you are going through this trial-and-error process, you constantly rank the critical elements in order of criticality. Eventually, it falls into place. In the case of *Arabian Nightmares*, it fell into place within two weeks. The game worked, and this method works the same way whether you are using a historical event or a future that you have to write a projected history for.

This trial-and-error method is one of the main attractions of wargames. It's what I call the puzzle-solving attraction. You are initially interested for the historical situation. The game explains some of the historical situation, but it also tends to uncover even more questions than it answers. For example, a game involving tanks leaves the players wide latitude in determining what combination of tanks and other weapons will be the most effective. You analyze as much as you can, but the remaining questions can be answered only through play.

## STRATEGY AND TACTICS OF PLAY ✱✱

In wargames as in real life, strategy is largely dictated by the available tactics. If tactics are what you do in detail, strategy is what you do with your total resources. If your tactics are such that your forces are much more effective in rough and wooded terrain, then your strategy will dictate that you need far fewer forces to hold the wooded terrain and, if perchance you have specialized units that are more effective than the enemy in attacking in rough and wooded terrain, then much of your operations would take place on that kind of ground. That's a fairly blatant example of how strategy is dictated by the tactical abilities of your units. In wargames, tactics are also affected by certain mechanical elements common to most games. The chief among these is the hexagon grid itself.

As you can see from looking at *The Drive on Metz* (and it's a good idea to refer to the game map while reading this section), there's a best way and a worst way to defend on a hexagon grid. The best way is to have an empty hex between each unit, have your defending units in a straight line along the grain of the hex. The "grain" of the hex grid is a straight row of hexes (the hexes stacked on top of each other like barrels). The worst defense is one in which there is a unit in every hex along the grain of the hex. This prevents you from covering a very wide area, usually makes it easier for your units to be surrounded, and also ties down a large number of your forces. The next-worst defense is defending against the grain, since this allows the attacker to bring a minimum of three units to bear against each of your defending units. When you're defending against the grain, you also have the option of having a nonstraight line, since you will again have no more than three attacking units against each one of your defending units. This can actually be an advantage in that, in almost any situation, you won't be able to avail yourself of a straight line because of the way the terrain shows up in irregular patterns across the map. That hill you must defend or river crossing that must be held always seems to be in an awkward position. Thus, the straight line is the unnatural situation.

Another type of defense that is not really a defense at all is the one in which you have two empty hexes between each unit. This is actually not a defense but a screen, since the loss of one unit or even the movement of one unit one hex in the wrong direction opens up your line and allows your enemy to penetrate into your rear area and easily surround your units. A situation such as this assumes one of four things. One, you don't have sufficient forces to cover the front, in which case there's not much you can do about it. Two, you do have sufficient force to cover the front, but you prefer to have a strong reserve (in which case you know what you're

## FOR THE DEFENSE

In the *Blue & Gray* game system (and in almost all game systems having advance-after-combat and Zones of Control) there is a basic defensive doctrine which should be followed with virtually religious zeal. It is: the defensive line should be as straight as possible, running with the "grain" of the hex pattern, and there should be one vacant hex between each defensive position.

*OPTIMUM DEFENSIVE LINE*

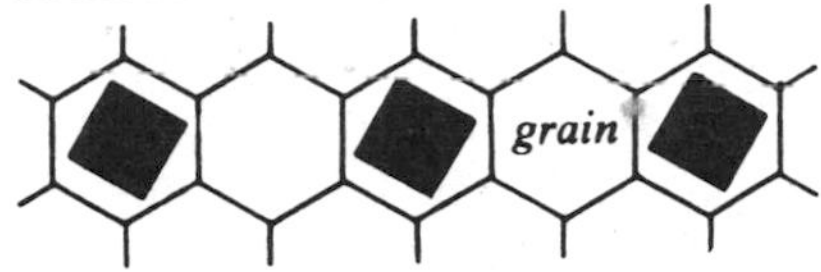


The advantages of such an arrangement are these:

1. No unit can be surrounded when a part of such a defense (neither in the Enemy Movement Phase nor by advance after combat).

2. There are only two hexes from which any position in such a line may be attacked.

3. The line is the most economical use of units consistent with good defensive coverage.

4. If an Enemy unit succeeds in advancing into one of the positions in the line, it is relatively easy to counterattack such an exposed unit and regain the line (there is also a good chance that the advanced Enemy unit can be surrounded and destroyed).

*THE LEAST DESIRABLE DEFENSE*

The least desirable defense is one in which there are two vacant hexes between each defensive position in the line. This type of line allows the units to be easily surrounded (by three or more attackers) and destroyed. Such a position is easily shattered and difficult to retake.

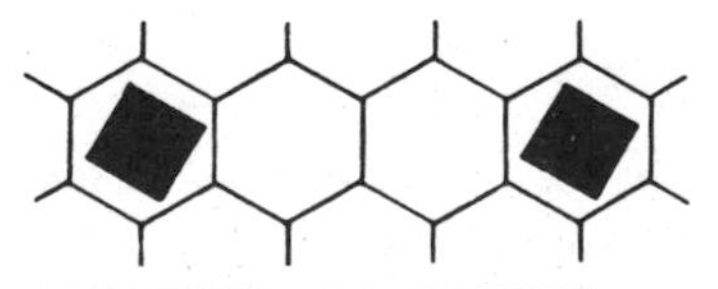

**Originally Published in MOVES magazine.**

*THE SECOND LEAST DESIRABLE DEFENSE*

It is also not desirable to defend in a "packed" line, i.e., one in which all of the defending units are adjacent to each other. This wastes units, and, more importantly, allows the attacker to make front-to-flank attacks (see main text).

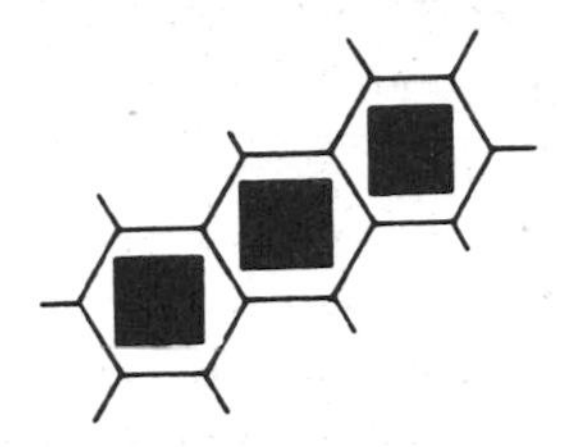

*DEFENDING AGAINST THE GRAIN*

When defending on a line situation against the grain of the hexagon grid, the units should be positioned in a manner similar to the optimum defense, i.e., with a single hex between each. Because of the geometry of the hex grid, the attacker will be able to attack any particular unit from three hexes.

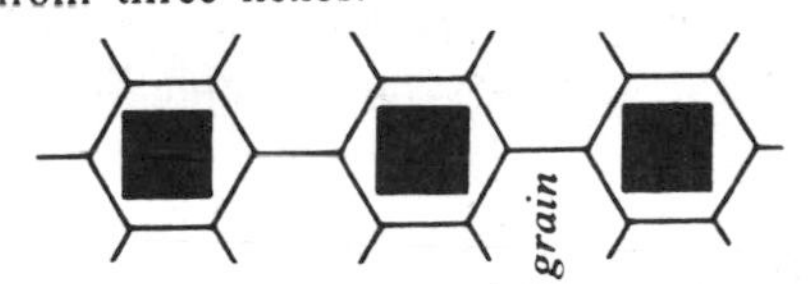


As an alternative to a simple straight line against the grain, the Player may wish to "refuse" the flanks of every other unit, making these forward positions the strongest unit or stack of units. This makes the line strongest in the areas where the Enemy can be the strongest (i.e., the only hexes against which three attacking groups can bear are those containing the strongest defending units).

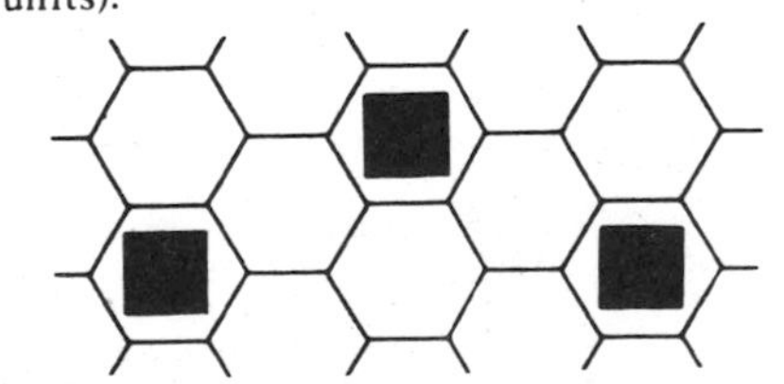

Players should note that these defensive formations are idealized and do not take into account local terrain variations which may force upon the Player less orthodox defensive deployments.

doing); and if the enemy does break through, you will presumably know what to do with your superior reserve. Three, you've just made a stupid, terrible mistake and are about to lose the game because you left two empty hexes between each defending unit. Four, you took a calculated risk that one particular part of the front would not be threatened and left this area thinly held. If you calculated correctly, you can recover from this, if not, go back to third item.

Another one of the standard features of hexagon games is the ability of the attacker to advance into the defeated defender's vacated hex. This is usually known as "advance after combat" and is an exception to the general rule that units may not move during the part of the game turn where they conduct combat operations. Remember, in most games you cannot move once adjacent to an enemy unit. On the face of it, that vacated hex is a new possession of the victorious attacker, but it can be a poisoned gift if on either side of it are additional defending units. Therefore, you must observe a couple of very important rules when deciding whether or not your attacking units should advance. There are basically three reasons why you would want to advance. First, this would normally pin down additional defending units. You may wish to do this even at the risk of losing the advancing unit. Second, as a corollary of the first reason, you may wish to set up a defending unit for elimination by surrounding it with flank attacks and then destroying the unit just surrounded (and thus unable to retreat).

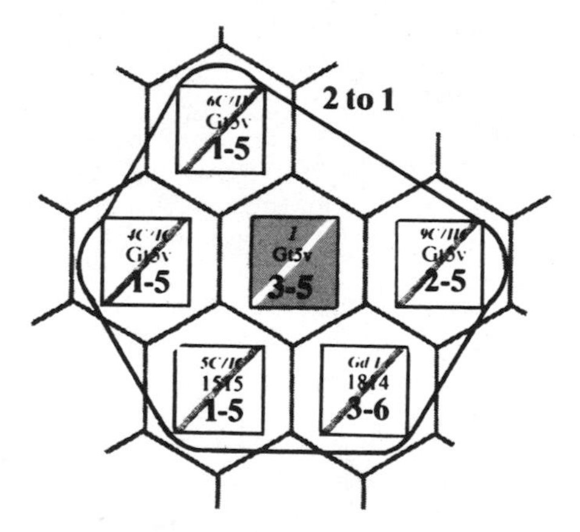


**Surround Attacks**

The third reason is for you to be in favorable terrain. This is often the case in situations in which you are fighting over possession of a town or another key terrain feature. Since, in many games, you are forced to have combat if you are in an enemy unit's Zone of Control (one of the unit's adjacent hexagons, which limit the movement of enemy units entering it), by your taking the favorable terrain, the enemy will then be forced to attack you at unfavorable odds.

Again, the only big risk in advancing after combat is that the advancing unit would put itself in a position to be surrounded and destroyed in the next turn. This risk must be weighed against the potential advantages.

The defender will also have opportunities occasionally to advance after combat if the attacker gets an unfavorable result and must himself either retreat or be destroyed. Normally, the defending unit will not advance after such a favorable combat. In the first place, since the defender is going to move next anyway, there's no advantage to be gained by advancing in order to pin enemy units, although there are exceptions. There could be a situation in which the defending unit is already immobilized by enemy units all around it and, in its own turn, would not be able to move anyway. In a case such as this, your "lost battalion" might as well do all the damage it can, since its days are numbered. In general, the only reason for the defending unit advancing after combat is to disrupt the attacking side's operations. But again, you must weigh the possible loss of that unit against any possible advantage. When dealing with combat operations on the hex grid, you will quickly discover that when playing another person, the instinct for self-preservation rapidly asserts itself in the form of an unbreakable line of your units facing an unbreakable line of your opponent's units. Victory, then, usually boils down to how best to break the line.

I should point out that this is a classic problem for game designers. Granted, there are many situations that are basically lines of opposing troops staring at, and sparring with, one another, unable to do much damage. But what makes historical wargames so interesting is that most battles did have a winner (war, being a game without rules, does not lend itself to draw-type situations). These victories were usually the result of any number of factors that, if the designer can incorporate them successfully into the game, will make for a rich and varied situation.

Some of these factors had to do with the way the armies were set up at the beginning of the battle. Others pertained to restrictions (often self-imposed) placed on one side or the other with regard to moving or using their combat forces. Other odd but interesting effects come from the historical victory conditions, which often have nothing to do with logic but are nonetheless critical because at the time they seemed like the right things to do. For example, in the *Drive on Metz* game, the Americans would have been much better served by trying to get around Metz from the north instead of from the south. At the time, going around from the south seemed a great idea, but, with the advantages of hindsight, we can see that the Germans north of Metz were weaker than we assumed. History provides a never-ending procession of examples such as this. Without such odd and illogical behavior, history would be dull indeed, and there wouldn't be much need for games such as ours.

All right, but let us assume logic does prevail, and you're faced with a line of units. You've got to crack it. If there isn't any flank you can get around, and assuming it has to be a frontal assault, pay close attention to the following points on how to do it.

First of all, you must be the superior player. That is, you must have more strength on the map than the other player. To do this, you must first calculate the maximum of combat strength you can bring to bear in any one hex. This means analyzing your units and, if you have stacking of units (more than one unit on a hex) in the game, you must take that into account to see how strong a "shock" force you can come up with. Then you must calculate the maximum defensive strength of the various enemy positions. What you are trying to achieve is good combat odds, normally 3 to 1 or better, so that you can at least push back defending enemy units. What you are going to try to do is compromise the defender's entire position by taking key defensive positions away from him. As this is done, you will force the defender to do one of two things: either retreat his entire line, which will be costly since you will have already pinned down (moved adjacent to) some of his units, thus forcing him to sacrifice the pinned units in order to retreat the others; or, if he decides to stay where he is, you will create a bulge in his line that you will eventually surround and destroy. This is much like a siege, in that it's going to require a lot of carefully planned attacking. It will also require a bit of luck, since if the defender is on the ball, he is going to see what's coming and form his own shock force for the purpose of taking yours on.

You now begin to see that technical expertise will take you only so far. A lot of it is psychological. Much of it depends on the other player making a mistake: something as simple as putting one unit in the wrong hex, in a position that looks okay at first, but on further analysis is not okay at all. In this respect, playing a wargame is much like playing chess. There are strong similarities, and the last time we checked, more than 98 percent of wargamers had played chess. To be successful with wargames, you've got to plan, look ahead, and be lucky. All of this advice also works when playing solitaire. It's an amazing experience to take advantage of, and be tripped up by, your own mistakes on both sides. Kind of makes you humble.

## THE IMPORTANCE OF QUALITY OF PLAY ✱✱

We have discussed some of the basic elements in playing wargames. Ultimately, however, it is the quality of play that will determine who will win. Look at it this way. Two players of equally high skill in tactics, technique, and analysis will play a dull game, the outcome of which will largely be determined by the intrinsic bias of the game for one side or the other. This rarely happens. In chess there is often reference to "dull" games in which both players did what was expected. In wargames the situation is far more ripe for innovation and surprises.

It is relatively simple for a player to develop a good (even superior) tactical skill at games. What is more difficult is acquiring superior technique and analysis ability.

Technique is a grab bag of skills involving the evaluation of the elements of the game and the options thus available. Take, for example, our *Drive on Metz*. The German player has considerable variety in his forces. He has strong units and weak units, fast units and slow units. Because this is a simple game, there are none of what we call "special" units with unique abilities. However, most games do have such units. In *The Drive on Metz*, for example, there could be artillery units that could project their combat strength over three or four hexes.

The German player, in order to maximize his chances for success, must mentally (or on paper) draw up a list of his units ranked in order of their abilities. The easiest way to do this is simply to multiply the unit's attack value by its movement value. The resulting number would be the unit's total value. Line up all the units on the list by value. You now have what amounts to a list of your best and your worst units. Naturally enough, your best unit will be a relatively strong one that is very fast. On the other hand, units that are relatively strong and slow are also useful, but in different ways.

The German player must then analyze the terrain that he needs to defend. This can be done by dividing his front into sectors—say, blocks four or five hexes long and three or four deep. Assign each type of hexagon a value according to its defensibility. An open- or clear-terrain hex would get a value of 1. A fortified hex would get a value of, say, 3 or 4, and the woods and town hexes would get something in between. Thus, each sector would come up with a number indicating its relative defensibility. This defensibility is almost as good as having additional units, since any unit that is placed in there is multiplied in its defense value. At this point, the player can then proceed to evaluate what the enemy forces can do, rank the enemy forces, and, in effect, think like the enemy. This is probably one of the key elements of good chess play, looking at the game from the other player's point of view.

Once new players realize the value of this analysis, the quality of their play increases enormously. While these techniques will improve your skill at the games, don't forget the opportunity just to sit down and start playing. Most gamers do this quite a lot. While the games lend themselves to some very interesting analysis, it's also entertaining and educational to just, as chess players put it, "push wood."

One thing you will notice is that the analysis of wargames is not the dry, systematic, mathematical (which, admittedly, some people enjoy) analysis that you have in chess or other abstract games. Rather, you're analyzing a real historical situation. Because the games demand such realism and accuracy in their historical information, players get the feeling that they are participating in a study of human events, which is exactly what they are doing. This makes the analytical work more palatable.

So you study the hell out of the game. In a tactical sense, this is a pretty good approach. It still doesn't mean you're going to do very well unless you show some ability at analysis. By analysis I mean what you can do with all of this in the game itself. For example, so what if you outnumber the other fellow two to one? If he's holding the mountains that you have to get into, this might actually be putting you at a disadvantage. Analysis means determining what type of specific tactical situations will arise in the game and what is your optimal reaction to these situations. Your tactical skill and your techniques are merely the tools that you apply on the basis of your analysis. Many gamers go back to historical accounts of the battle they are gaming. They often find good analysis there by the original participants. If nothing else, you get an idea of what not to do.

Between two really skillful players, it is the analyses that will cause most of the excitement. Often both players will correctly analyze the situations in a general sense. Where they may differ is in how they apply the tactics or techniques to implement their analysis. This makes for very exciting play. For example, one player may have determined that in order to win a game of *Waterloo*, he must attack certain British units during a specific turn of the game with a certain force. This force can either be all ground forces or it can employ a large amount of his artillery. How much of his artillery he can spare will largely be determined by what he is doing during the entire battle that turn. Thus, while the player has good tactics for using the artillery and good technique in analyzing where it will usually be most needed, there is still enough room left in any wargaming situation to go one way or the other. There are so many hundreds of unconscious decisions being made when any one of these situations is being analyzed that it is a truly creative act when one player comes up with an analysis of the particular turn of the game that is demonstrably superior to another player's analysis.

While I understand these techniques and have used them to one extent or another, I am not the greatest practitioner in the world. My knack is for designing the games, not playing them. When I do play, I get distracted by my habitual tendency to try to figure out how the game was put together and then to try and come up with better ways to do it. Meanwhile, my opponent is usually just playing the game. Many times I've been on the receiving end of much better users of these playing techniques, and I find it much more pleasurable to watch two other players knock each other's brains out in a creative contest of wargaming tactics, technique, and analysis.

## WARGAMING TECHNIQUE ✱✱

Wargaming techniques are methods that are generally applicable to any type of game. They are just more applicable to wargames because wargames have more things going on in them. There are more elements to apply technique to.

As described above, one of your basic techniques is analyzing your units. Once you have analyzed them, you then have a second technique known as "effective deployments."

Because each side has a wide variety of different kinds of forces, the positioning of these forces helps determine their effectiveness. For example, your lighter and slower units are generally garbage and are best put in the front line where they can absorb the enemy's initial blow. Your weak but fast units are often best deployed on flanks, where they can rush about or at least slow down any enemy attempt to surround you. Your strong but slow units are generally best deployed right behind your weak front-line unit so as to move up and stop any attempt at pushing into your position. Your strong and fast units are best used as your general reserve to be kept in the middle of your position so that they can rapidly be shifted anywhere either to defend a key area or to exploit a successful attack. Range-fire units (units such as artillery or archers, etc., that can "fire" their combat power over two or more hexes) are usually deployed to make maximum use of their ability to project their combat power over a wide area. Thus, missile units can reinforce nonmissile units. Needless to say, when you're analyzing a unit's effectiveness, a missile unit becomes important far beyond its basic combat and movement ability.

For any other specialized unit, the same applies. For example, some games have leader units, which are necessary to get the combat units to move, fight, or do just about anything. These units must be deployed so as to enable you to do what you want to do. Put them in the wrong place, and you're going to be high and dry when it comes to getting things moving. Note that the above advice on using various types of units also applies to chess, and has been the accepted wisdom in military circles for several thousand years.

Just to sit down and attempt to develop different (and one would hope better) deployments for a particular game is an interesting exercise that many players indulge in. This type of activity often leads to interesting surprises, as when the player who has done his homework unveils a new and more effective deployment on an unsuspecting opponent.

Many players prefer to tinker with the game in this manner rather than actually play it. Tinkering is a low-hassle, mentally stimulating, take-things-as-they-come activity. Playing the game tends to be a bit more intensive even if you're only playing against yourself. Most gamers tend to be people who like to tinker with things, and games are eminently "tinkerable."

## TEMPO AND SHAPING ✷✷

Either player in the game has the option of speeding up the level of events. In most games, each player takes his turn, first moving and then attacking, and one player taking an inordinate amount of time doing so. But the other player can still unilaterally increase the activity level of the game by moving more units and engaging in more combat, since this forces the first player to respond.

It is inherently easier to increase the tempo of some games, primarily the ones with relatively bloodless Combat Results Tables. This means that the risk to an attacker is considerably less than in a game in which each combat produces some loss for somebody. You can still increase the tempo in a game with a deadly CRT. It simply takes more skill to do it without automatically losing too often. Once you increase the tempo, you gain an immediate advantage, because the other player (assuming you have somewhat equal skills) doesn't know what's going on. He can only assume that you're up to something. If you're smart, you will be up to something and not just raising a little hell. As the tempo of play increases, as more attacks are made, as more units in contact move to and fro, the tactical situation is constantly changing.

At any given time, you must, if you are on the ball, be aware of the tactical possibilities of all of your units. Who is in danger, who has opportunities sitting in front of him, where the opportunities or potential opportunities happen to be. As you increase the tempo of play, you are changing all of this rapidly from turn to turn. What you are doing here to a large extent falls into the area of gaining a psychological advantage.

This is no different from any other human-conflict situation. In fact, let's not confine it to humans, let's take animals in general. There's always some sort of pecking order instantly established in animal groups. One animal or the other will soon gain some sort of advantage. This can often work to the "advantaged" side's disadvantage. Some players, for example, allow themselves to assume the role of the "lesser player" and thus lull the "superior" player into a false sense of superiority until such time as the "lesser" player lowers the boom.

Many players also have an overly slavish approach to what happened historically. If one side, for instance, was the defender historically, the player for that side assumes that in playing the game on that event he must adopt a defensive attitude. Quite often this is just the opposite of what is the best for him in that situation. Again, a perfect example is any Waterloo game (there are many). In that situation, the Battle of Waterloo in June 1815, the French were indeed attacking the British. But the British were not at that much of a disadvantage. They were capable of counterattacking, and counterattack they originally did. In the game, you can benefit by being even more aggressive than the British originally were. This is especially true if you are playing against a sloppy French commander. I have seen many players become victims of their own overly literal reading of the historical accounts. If faced with an inept Napoleon, take advantage of the situation.

Shaping is a technique in which you use things such as tempo, psychological advantage, tactics, technique, and analysis to put the enemy in the position you want him. This is a legitimate technique whether you are on the offense or the defense. What it comes down to is that everything you do in the game is done with a specific objective in mind. You don't advance in a certain direction, you don't retreat in a certain direction, unless this fits in with the plan you have devised. Many people are afraid of planning. They simply react, and this gives the other player an enormous advantage. Unless the other player is also reacting, which then produces a muddled and confused game quite similar to many historical battles fought with similar reactive attitudes by the original commanders.

To react without planning is to let the other player shape your game. When you get two players who are reacting, it becomes a very interesting and unpredictable game, and many people prefer to play it this way. There is nothing wrong with that. However, if you want to obtain more control over the game and do things in a more systematic manner, you must come up with a plan, with objectives.

The most obvious objectives are fulfilling the game's victory conditions. Yet a plan consists of more than saying, "Well, these are the victory conditions, and this is how I'm going to do it." You must work things out in more detail. For example, in *The Drive on Metz*, the German player must determine which units he is going to put where to defend what. Up in the north by Thionville, he has to worry about the Americans putting a main thrust where there are so few German units. Thus, he must make some of his more mobile units available to be sent north very quickly. In the center, where the Germans are probably most vulnerable, he has the advantage of terrain and fortification. But the fortifications, once lost, are almost impossible to retake. So particular care must be taken with regard to what is put in the center to hold these key areas.

Developing the ability to do this, to shape one's play, takes time. All I am pointing out here is that the opportunity exists. Many players will find that they do not care to do that much work to play a game. However, I think most players would benefit from at least giving it a try, because they would probably find, at the very least, a simpler version of some of the techniques I'm talking about here, which they can regularly apply without any additional expenditure of energy. The object of all of this is not to turn the games into a lot of drudgery, but simply to point out things that players will probably eventually stumble upon themselves anyway. Many of the techniques we are talking about actually make the games more enjoyable. You feel less "lost," and you feel more in command of the situation (no pun intended).

Most of the techniques dealt with above are applicable primarily to two-player games as well as the individual-player activities of multi-player games. However, multi-player games do have a dynamic of their own that is not found in two-player games. When you get three or more people involved, there is the possibility of coalition. Also, funny things happen in people's heads as they comprehend the fact that they are dealing with not one person but two or more. This usually has some interesting, predictable results.

For example, in the early 1970s I did a game for Avalon Hill called *The Origins of World War II*. This game was a multi-player game in which players assumed the roles of the various powers in Europe. Although the players were told that they had to stop Germany if they wanted to prevent World War II, each player had his own specific victory condition, and invariably everybody would look out for himself, and Germany would walk into all the adjacent countries and would usually end up winning the game anyway.

In *Dungeons and Dragons* and most other role-playing games, a similar situation exists. It is understood, and it is often made explicit, that the players will gain more individually as well as a group if they cooperate with one another. Invariably, though, players will be more interested in looking out for their own personal gain at the expense of any possible cooperation or planning ahead with other members of the team. I can offer no pat solutions to this. It seems to be an intrinsic element in human nature. People like to "do it themselves." This is one of the chief attractions of a role-playing game in which you are playing a role. In fact, these games might more appropriately be called role-building games. You are building a role, and by playing it out, you give it substance. Cooperating with somebody else somewhat diminishes the individual in a game such as this.

It is a bit more blatantly obvious in role-playing games, but it is an element present in any multi-player game. People cooperate only briefly and only to confront and solve immediate problems. This gives multi-player games a great deal of flexibility. They can go on and on because no one person is going to gain a long-lasting advantage. It's more of a problem in role-playing games in which you are not really playing against the other players, but while this attitude exasperates the dungeon masters (or "gamesmasters," who are running the game), the players don't seem to mind that much and take their licks like the heroes they are trying to be.

## HOW TO PLAY WITHOUT AN OPPONENT ✷✷

Playing wargames solitaire is by far the favorite mode for most wargamers. The most common reasons for playing solitaire are lack of an opponent or preference to play without an opponent, so that the player may exercise his own ideas about how either side in the game should be played without interference from another player. Wargames are, to a very large extent, a means of conducting historical experiments. So another player just gets in the way at most times. For those players who do like to play with opponents, solitaire play is valued as a means of perfecting tactics and techniques in a particular game that will enhance the chances of success.

There are any number of techniques, which I'll get into later, for making a game a true solitaire game. Such a game is one in which an opponent is simulated by the game rules. But by far the easiest and most rewarding approach is to simply play each side as if it were the only side you were playing. This is not as difficult as it sounds. Granted, there will be some bias, but not be as much as you might think. On each turn you either turn the map around and play the other side, or go to the other side of the table, or simply place the map so that you're in between either side's forces and move accordingly. It is sometimes interesting to go to the other side of the table so that you are behind the forces that you are playing. It changes your perspective on the game considerably. Simple things such as this will provide you with a rewarding solitaire game.

For those of you who care to do a bit more work, there is another approach, which requires more effort but has had its proponents over the years. In this method, you sit down with the game and analyze the situation each side is faced with, and prepare a plan. This plan organizes each side's forces into half a dozen or so larger organizations. In other words, say in a game one side had 40 playing pieces on the map, each representing a battalion. The player would organize these battalions into divisions of five or six battalions each. The organization should follow whatever logic the battalion's deployment on or entry onto the battlefield dictated. Then, consulting the victory conditions and the number of turns in the game, the player would set up a timetable of what he felt reasonable goals should be for each of these divisions. He would also set loss levels at which the plans for these divisions would have to be changed. The same technique would then be applied to the forces on the other side. These plans would then be executed by the player, who would simply be operating at a lower level than the theoretical commander of each army. This is merely a more formal (and to some people probably more comfortable) approach to my method of turning the map around and playing the other side.

There is no mystery in playing solitaire. Most players do it, and many of them find it rewarding, simply by not getting wrapped up in any emotional desire for one side or the other to win. It is probably true that the older gamer has an easier time doing this, but then there's "comfort" to be taken in the fact that we all grow older, and perhaps this relaxed attitude comes a little sooner to people who play wargames.

## PLAYING BY MAIL AND TEAM PLAY ✷✷

Chess by mail has been around for years. It certainly changes the style and tempo of the game. The same is true for playing wargames by mail. The main requirement is that you have a place where you can leave the games set up, although this is not absolutely necessary, since playing by mail requires that you record the location of each playing piece each turn. For this reason, most of the games played by mail are fairly small ones with a small number of playing pieces on the map at any one time. Thus, it would be possible to put the game away after each turn. Most players, however, tend to leave them set up someplace on a piece of plywood or something so they can be stuck off somewhere out of harm's way. Some gamers even build racks with shelves on them (24 by 36 inches per board) and store a dozen or more games "in play" at any given time. To find opponents, we suggest you get in touch with AHIKS (Avalon Hill International Kriegspiel Society, and it *is* international), a long-term organization of play-by-mail players, or the Avalon Hill *General,* which provides free ads for subscribers for the purposes of getting by-mail opponents. For AHIKS's current address, write the Avalon Hill Game Company, and they'll probably be able to help you out.

Oh, by the way, about 2 percent of wargamers (according to several decades of surveys) indulge in play by mail, but these tend to be a very active and interesting 2 percent. It's a group worth getting involved with, especially now that it's growing through the use of computers to transmit game moves and other information.

Twenty years ago, wargaming and gamers were rather young. There weren't too many gamers who owned a house or had the room to set up any game, especially a large one, for any length of time and just leave it there. Over the past 20 years, a more and more common phenomenon has been the emergence of small groups of gamers, usually no more than half a dozen, among whom one has a house with a room large enough to put a game in and leave it set up. The gamer group then gets together, usually about once a week for three to five hours, and plays the game. Depending upon the game, they might get through four or five turns or perhaps only one. They will play in teams. This is particularly attractive with the larger games that lend themselves to teams on each side. Playing this way, there is no pressure to finish the game, and under these conditions the large games really come into their own and can be quite enjoyable as you manipulate such a large simulation at your own pace.

This type of team play was also used during the development of these larger games. While working on one of the first multi-map games (*War in the East*: three standard 22-by-34-inch maps and hundreds of playing pieces), we had only four players, and we didn't want to keep two of them idle while the other two made their moves. So we divided the eastern front in Russia into four sectors. Each player was given a sector on each side. In other words, each player had a Russian sector and a German sector. The sectors were set up in such a way that no player was facing "himself" on the other side. We then began play. As soon as people would finish playing their Russian side (they might have to wait for the other "Russian" players to finish), they would switch over and immediately begin playing the German side. Play went on without interruption. Granted, the tempo of play intensified a bit for the individual since he no longer could depend upon the "break" when the "other side" made its move. Generally, every hour or so we would take a 10- or 15-minute break just to stop and stare at what we had done. Meanwhile, there were usually opportunities for shorter breaks, as not all players on a side finished at once.

This form of team play in general has many advantages. It brings players together on a social basis. It does not pit them together on a one-on-one contest (which often has some unpleasant competitive aspects to it) and allows the larger games (which continue in popularity despite their size) to be played to a conclusion. The results are, I have found, a compelling advantage to be able to watch a large game unfold week by week. It's quite an experience.

## HOW TO GET INTO GAMING PAINLESSLY ✱✱

If gaming is new to you, you have a problem in that there are so many games available, you may want to be able to play games on particular periods of history at levels of conflict (tactical, strategic, air, naval) without taking the chance of running into some game that requires a bit more seasoning than you might have at the moment. The problem is easy enough to solve with a little informed advice. First of all, if you haven't played a game before, you've got the fairly simple one in this book. Having mastered an introductory game, you are ready to go after dozens of games from various periods using the same system.

By a game "system," I mean a group of games using the same movement rules, the same combat rules, and often very similar Combat Results Tables and Terrain Effects Charts. The battlefields and the units of these games will be considerably different, to reflect the different situations. And there will be a few special rules for each of these additional games. Publishers have developed hundreds of games based upon the system used in *The Drive on Metz*. This game system is not only the simplest available but also the most commonly used, having been developed in the late 1950s for the first commercial wargames. Games using this system, with numerous variations, are available in every era and are a good way to start, as they enable you to go into whatever period you like without learning a lot of new game mechanics. Once you have mastered some of these simpler games (and each of them has additional little wrinkles which will expand your repertoire of game rules), you are ready to go after more complex games.

Some of the more readily available simple games are known as the "Avalon Hill Classics." These are some of the earliest games published. Some of them are over 30 years old and still in print. They are very similar to the other publishers' simpler games. Nearly all of the easy-to-play games pay homage to these early Avalon Hill games. The simple mechanics of these early games worked so well that no one has really been able to come up with a better approach for games that are easy to learn and play. The games comprised in the Avalon Hill Classics are *Waterloo, Stalingrad, Battle of the Bulge, Afrika Korps,* and *D-Day*. Other publishers have brought out simple games over the years, and there are always new ones available. Inquire at a game store or from one of the companies that sells games by mail.

## PLAYING COMPUTER WARGAMES ✱✱

All of the above has assumed that you would be playing a manual (paper) wargame. The same techniques apply to computer wargames, with a few variations and additional features.

Computer wargames have a number of notable differences from their paper predecessors. The principal differences are:

**Easier Rules—** Unlike manual games, which require that the players read and digest long, intricate booklets of rules, computer wargames have their rules embedded in the program itself. There is still a lot to be learned in advance, as the computer wargame is like any other computer program, and you still have to learn what pushing each button will do. But more and more computer wargames have their documentation embedded in the program. Once you learn what key (or icon) brings up the ''Help'' information, you can dispense with the printed instructions altogether. There is a great deal of variation in how easy computer wargames are to use. But the trend over the last 10 years has been for them to get easier, and they have always been easier to learn than their paper equivalents.

**Faster Start-up—** The bane of paper wargames has always been the tedious process of laying out the map and setting up the playing pieces before you could even begin play. A major advantage of computer wargames is that, in most cases, the ''playing pieces'' are already where they should be when you start the game. Those games that allow ''free set-up'' are still a problem, as you then have to use the keyboard or mouse to place the units. In general the technology in computer wargames is making this set-up problem easier with each passing year.

**Hidden Information—** A major advantage of computer wargames is that you can easily experience the ''hidden information'' aspect of military campaigns. It's not for nothing that spies are shot in wartime. Secrecy is often a matter of life and death, and the ability of the computer to keep you in the dark most of the time accurately re-creates this aspect of warfare. On the downside, this makes pre-play analysis of the game more difficult than with manual games.

**Saving the Game—** Besides the hassle of setting up the game, manual wargames are also wretched when you want to put aside a game to finish later. The map is a large object, and all those playing pieces can be easily displaced. Computers suffer no such problems. Press the right key, and your game in progress is saved to a disk. You can take it up again whenever you want.

**Superclerk—** A computer is very good at computing and keeping track of things. Manual wargames always suffered when burdened with too much record keeping. Yet data, and calculation, are what make a wargame different from chess and capable of re-creating history. Computers can excel in this department, and often do.

**The Little Picture—** One major advantage of paper wargames is that you can see everything. The map is big, but it's all there in front of you. Computers only have that little screen, which is often seen (literally) as a "window" looking at a larger map. You have to scroll around the larger map, although many computer wargames have the option of popping up a "strategic map" that shows key features of the entire playing area. As computer displays become capable of higher resolution (which is cheaper to achieve than larger displays), this becomes less of a problem. But the paper games still have an edge.

**Black-Box Syndrome—** Another advantage of paper games is that you know why things are happening a certain way in the game. All the rules and probability tables are right there in front of you. Yes, it takes a lot of effort to wade through all that detail, but you do end up with a good idea of how the inner workings of the game function. A popular benefit of this is the opportunity to change the game's rules and probability tables. Many players do this, and that's how gamers eventually turn into game designers. A computer wargame shows you very little of how it does its thing. The computer program just does it, leaving you sometimes muttering about mysterious "black boxes." Naturally, you can't change the program either. This inaccessibility puts off a lot of gamers who started out on manual wargames. But increasingly, new wargamers have seen only computer wargames. Some of these new gamers still admit to loss when they realize how much more enjoyable the game would be if they knew what was going on inside it and could make their own changes. Computer-wargame designers have become aware of this, and increasingly computer wargames have options to modify their procedures and also let the player know what's going on inside the black box.

**Paper Clones and Silicon Masterpieces—** The earliest (and some of the current) computer wargames were basically clones of manual wargames. Over the years, more and more computer wargames have concentrated on the computer's strengths and less on trying to clone a paper wargame. There are more and more silicon masterpieces being produced as computer-wargames designers find more ways to take advantage of the computer's unique properties. Paper wargames are different, in some ways superior, and don't clone well.

**Artificial Intelligence—** Most computer wargames are essentially solitaire games, with the computer playing the other side. Artificial Intelligence (AI) routines are used to provide the computer with playing smarts. These AI routines have got increasingly powerful over the years, although you will still encounter games with brain-damaged AI. On average, though, AI routines are still a cut or two below what the average human player is capable of. In general, the AI routines take a very calculated approach to playing the game. This is understandable, as a computer is designed for calculation. AI routines usually have problems with strategy and are not as quick to catch on to changes in the big picture. This said, there are two things you have to watch out for. First, some AI systems cheat. It's a dirty little secret of computer-game design that some games will take a shortcut and let themselves know all about your situation while keeping much of their situation hidden from your view. Often the player suspects this is the case, and often the net effect is simply to make the computer a more formidable opponent. Many players, however, are irked at this. Second, the AI often uses a random approach to selecting its next move. This is an easy way to make the AI opponent appear more formidable. Actually, if the AI is not all that predictable, it *is* more formidable. So, when you think the computer opponent outthought you, it actually just outguessed you. Unpredictability has always been considered a very human trait, so it shouldn't be surprising that the computer would adopt that trait and use it on you.

One of the biggest areas of new technology in computer wargames has been the introduction of more powerful Artificial Intelligence. You often have a number of choices in a game with regard to AI. In games that have achieved very powerful AI (that is, a human can't beat it), you are given the choice of a less capable AI routine to play against. Some games also offer you the option of different AI "personalities" to play against. Typically, you will be offered AI opponents who are "Crafty," "Cautious," "Daring," "Calculating," "Reckless," "Idiots," or "Completely Deranged." Something for everyone.

**Modem Play—** In the late 1980s, the ownership of modems (which allow computers to communicate with each other via the telephone) by PC users reached a critical mass. Millions of regular modem users began to wonder why they couldn't play their computer wargames with other gamers via modems. Publishers quickly responded, and soon several wargames were being published annually that supported modem play. At this time, it's almost mandatory that a computer wargame have "modem capability." Use of the modem in suitably programmed wargames is simple. Both gamers load the same game and put the game in "modem play" mode, one player enters the other gamer's phone number, calls that player via his modem, the two computers connect, and both players can play each other. Some of the nationwide computer services allow such gamers to save most of the long-distance charges by calling the service's local number to play other gamers worldwide. Players can once again play other gamers, now with all the advantages of computer wargames.

There are several variations in modem play. The simplest is to take a modem-play game and simply run a cable between the players of two computers in the same room. Another, older variation is to allow players to transmit their moves as a file. Each player reads this file to see what the other player has done, and then responds by creating a file of his moves to transmit back. Sort of like playing chess by mail. In that vein, many players are using electronic bulletin-board systems or networks like GEnie, the General Electric Computer Network, to play games on-line. It's interesting to note that while the earliest computer wargames could be played by two players, this never caught on. This required the players to take turns sitting down at one computer, and players wanted their own computers. Another aspect of human nature that was not discovered until PCs came along.

**The Spectator Sport—** Computer wargames with AI playing both sides become wargame simulators and, for the gamers, a spectator sport. In the beginning, computers weren't powerful enough to allow more than one side to be computer-controlled (and usually only one side; you had no choice of which side to play). Then, as machines increased in power, it was possible to put in programming that could play either side. At this point, it was no great leap to realize that you could let the computer play itself. This has become more popular, and it's basically an extension of the idea of using the original paper wargames as a means of analysis. Computer wargames that have a lot of options and scenarios enable users to mix and match options and let the computer play them out. Ironically, this is what the military (professional) wargamers have been doing for decades. Now you can buy very similar software for $50 or so. That's several million dollars less than the government pays. But then, the government buys only a few copies, not tens of thousands.

## THE TECHNICAL TERMS OF WARGAMING ✱✱

Many special words or meanings of words have been developed to assist in the play of wargames. Very few of these terms are unique to wargaming, and many of them are borrowed from scientific and military usages. The following list gives the more common terms you will encounter in playing historical-simulation games.

• **Abstraction.** A key concept in wargames whereby complex procedures in a historical event are much simplified in a wargame of the same event. This usually works, but sometimes it doesn't.

• **Advance (After Combat).** Usually, the attacker's unit ability to move into the space formerly occupied by the defending unit (which was forced to retreat). Sometimes an advance allows the successful unit to move more than one space.

• **Administrative Units.** Game counters that represent noncombat units that are still vital to the play of the game. Supply, headquarters (to control other units), and engineer (to build and maintain things) units are typical of this type.

• **Adventure game.** What some wargames are called, and what some games that are not quite wargames actually are. Adventure games are similar to role-playing games, especially those on computers, that involve one person (the player) solving puzzles and fighting opponents on an adventure. Often the military aspect is quite detailed.

• **Air Attack.** Aircraft attacking something. Often called "ground attack," as it is often aircraft attacking ground units. "Surface attack" is sometimes used to describe air units attacking ships.

• **Air Missions.** Games that deal with air power in detail often have to cope with the many different things aircraft can do. Typical missions are Anti-Air (or "air superiority" or CAP—Combat Air Patrol), Surface Strike (or "ground attack"), Air Transport, and Ferry missions.

• **Air Units and Combat.** Playing pieces are often used to represent aircraft units. Because the typical game is only two-dimensional, the player must keep clear in his mind that although the air units are being placed on the same playing surface as the ground units, they are not operating on the same level. Because of the special symbols used on the air units (usually they depict the silhouette of an aircraft), it is not too difficult to get used to this potentially confusing graphic situation. In addition to air units attacking one another, you also have ground units that have as one of their functions the ability to attack air units and in effect defend friendly units on the ground.

• **Airlift.** In addition to specialized units capable of combat in the air, there are often units capable of transporting ground units by air. The ability of these units is often referred to as the "airlift capacity" and usually denotes a specific number of units that can be moved in a given game turn or period of time. The airlift capacity sometimes is expressed in terms of airlift points, which are often equivalent to the transported units' combat strength or stacking points.

• **Air Range.** This is normally the number of spaces on the map that air units may move in their turn. Since air units are moving above the ground and not on it, they usually ignore all terrain restrictions and generally have much higher movement allowances than ground units.

• **Airborne, Airlanding, and Airmobile Units.** An *airborne* unit is normally considered one dropped by parachute. These units undergo a chance of being destroyed or otherwise incapacitated because of the dangerous techniques they use in getting on the ground. *Airlanding* units are units that are landed normally on an airfield in or very close to enemy territory. An *airmobile* unit is a unit that is landed by helicopter.

• **Ammunition.** This is normally represented in the game, when it is distinctly represented at all, as an item of supply without which units cannot attack. The amount of ammunition you have is sometimes represented by ammunition points, each point representing the ability of, say, one unit to make one attack. This basic approach can become much more elaborate. Often, simpler ammunition rules simply provide a probability that you are out of ammunition.

• **Armored Unit.** A unit in which all of the forces are using armored fighting vehicles is considered an armored unit.

• **Assets.** A purely military term, referring to all combat and combat-support units as well as supplies and the like a commander has at his disposal. The term has entered the vocabulary of wargamers. It's usually the other way around, as wargamers do not generally use a lot of strictly military jargon.

• **Attack.** One unit (or a group of units) attacking a single defending enemy unit (or group of units). The player doing the attacking is referred to as "the attacker."

• **Attrition.** This term represents gradual loss of a unit's strength, rather than complete loss, which is often what happens in many board games. In the case of individual units, this is represented by one of three methods. One technique is to have a number of playing pieces representing each unit, with each of the different playing pieces for the same unit differing only in the combat-strength value printed on them. Another method is to place another playing piece with only a number printed on it underneath the piece representing the combat unit, the numbered playing piece representing the current combat value. As the combat unit loses strength, a numbered playing piece of lesser value is placed under it. A third method is to have a roster printed on a separate sheet of paper. Each unit would have next to it a row of boxes, each box representing one strength point. As the unit suffered attrition, the boxes would be checked off in order to show a lower strength. Computer games have a substantial advantage in that they can keep a very precise record of losses without burdening the player. Sometimes attrition is applied to one side's entire force, in which case (normally during an "attrition phase" in the game turn) a Probability Table is consulted to determine how many units from an entire army will be removed from the game to represent attrition. This type of attrition usually represents the losses caused by sickness, starvation, or disease.

• **Blast Radius and Strength.** This is generally a circular area represented by the number of hexagons from the hexagon of impact in which the bomb or shell hit. Also applies to just about any explosion (grenades, mines, and so on). The blast strength is the numerical rating of the effect of the explosion, and this declines as one gets farther away from the impact hexagon of the explosion.

• **Breakdown.** Often units can be broken down into two or more smaller units. Usually, the total strength of the two or more smaller units does not equal the combined strength of the original larger unit (because the sum is greater than the parts). A larger unit is usually broken down in order to cover a larger area. The term "breakdown" is less frequently used to describe a unit or equipment that doesn't work (the more common meaning outside of wargames).

• **Case.** This is a numbered (sometimes lettered) paragraph that states or explains a specific game rule. Cases are usually numbered, although sometimes letters are used. See *The Drive on Metz* (pages 132–143) rules for examples.

• **Casualties.** Historically, this term refers to the dead, wounded, and missing resulting from a battle. Casualties in a game, however, normally represent more than the actual personnel losses and include the functional breakdown of military units as effective combat forces due to dead, wounded, and missing troops as well as equipment losses and damage. It is for this reason that often the losing unit in a battle is removed from a game. Historically, as little as 30 percent of a combat unit's personnel being killed, wounded, or captured is often sufficient to render that unit totally ineffective as a combat force. Casualties may also result from attrition and friction.

• **Character.** In role-playing games (RPGs) each player (four to six is the most common number) represents a different person (some are non- or semi-human like dwarves, elves, and the like). Each person has different characteristics (speed, strength, intelligence, etc.). Some computer games, and wargames, use similar individual characteristics.

• **Close Assault.** A term used in tactical games indicating hand-to-hand fighting-type attacks made by infantry or infantry and combat engineers—the modern equivalent of "going after them with the bayonet." Note that there is actually very little hand-to-hand combat, although it is more common for troops actually to see whom they are shooting.

• **Conflict Simulation.** The more technically correct term for what we commonly refer to as "wargames." Wargames are actually only one subgroup of the much broader group of games that in theory includes all situations involving struggle between two or more opposing forces. A conflict simulation attempts to produce a playable but essentially mathematical model of such situations.

• **Combat.** This occurs when two or more opposing units interact with a result that produces losses, retreats, or other changes in the status of the opposing units. These actions usually take place during a combat phase, which is a distinct part of a game turn. As most games put it, "Combat occurs between adjacent opposing units. The units of the phasing player are used to attack and the units of the nonphasing player defend, regardless of the overall strategic situation."

• **Combat Strength.** A numerical rating of the unit's ability to attack and defend. This rating is expressed in terms of combat-strength points. In some games, a unit has one combat-strength value to be used both for attacking and defending. In other games, the units have two combat strengths, an attack strength for attacking and a defense strength for defending. These strengths are calculated on the basis of the raw firepower of the units and modified by qualitative factors such as training, leadership, organizational effectiveness, and experience. Calculating these combat strengths is one of the more interesting aspects of game research and design.

• **Combat Results Table (CRT).** A probability table that shows the possible results of all combats allowed within a particular game. The greater the ratio of attacker-to-defender strength, the higher the chance of success. Because so many things can go wrong during the combat itself, a die or other random-number generator is used to determine the actual result. These tables are usually calculated based on what information is available on actual historical losses.

• **Combat Results.** Once the die is rolled and the CRT consulted, the actual results of the combat will be quite specific. Given below is a list of some of the more frequently used combat results and their meaning:

*Ae (Attacking Units Eliminated).* All of the participating attacking units are destroyed (removed from play). This result is common when a very unfavorable attack is being made, one that has a high probability of the attacker being effectively wiped out.

*Ar (Attacking Units Retreat).* All of the participating attacking units must retreat one or more hexes. One of the more common results. Attacking is more difficult than defending, and attackers are frequently thrown back.

*Ad (Attacking Units Disrupted).* All of the participating units are

## [7.6] COMBAT RESULTS TABLE

| Die Roll | Probability Ratios (Odds) Attacker's Strength to Defender's Strength | | | | | | | | | | Die Roll |
|---|---|---|---|---|---|---|---|---|---|---|---|
| | 1-5 | 1-4 | 1-3 | 1-2 | 1-1 | 2-1 | 3-1 | 4-1 | 5-1 | 6-1 | |
| 1 | Ar | Ar | Dr | Dr | Dr | Dr | De | De | De | De | 1 |
| 2 | Ar | Ar | Ar | Dr | Dr | Dr | Dr | Dr | De | De | 2 |
| 3 | Ar | Ar | Ar | Ar | Dr | Dr | Dr | Dr | Dr | De | 3 |
| 4 | Ae | Ar | Ar | Ar | Ar | Dr | Dr | Dr | Dr | Dr | 4 |
| 5 | Ae | Ae | Ar | Ar | Ar | Ar | Dr | Ex | Ex | Dr | 5 |
| 6 | Ae | Ae | Ae | Ae | Ar | Ar | Ex | Ex | Ex | Ex | 6 |

Attacks executed at greater than 6-1 are treated as 6-1; attacks executed at worse than 1-5 are treated as 1-5.

**EXPLANATION OF COMBAT RESULTS**

Ae = **Attacker Eliminated.** All Attacking units are eliminated (remove from the map).

De = **Defender Eliminated.** All Defending units are eliminated.

Ex = **Exchange.** All **Defending** units are eliminated. The Attacking Player must eliminate **Attacking** units whose total, printed (face value) Combat Strength **at least** equals the total printed Combat Strengths of the eliminated Defending units. Only units which participated in a particular attack may be so eliminated.

Ar = **Attacker Retreats.** All Attacking units must retreat one hex (see 7.7).

Dr = **Defender Retreats.** All Defending units must retreat one hex.

disrupted. This usually means that all of the affected units are pinned in the hex they are in or otherwise prevented from attacking or performing any action they would normally be capable of. This is a variation on Ar (Attacking Units Retreat).

*A1, A2, A3, etc. (Attacking Units Lose Indicated Number of Steps or Strength Points).* One or all of the participating attacking units are reduced by the number indicated (or retreated the indicated number of hexes). Another variation on Ar (Attacking Units Retreat).

*Aex (Attacker Exchange).* The attacker loses participating units whose face value is at least equal to the face value of the defending force. Sometimes this is expressed as in AEx, meaning that the attacker loses participating units equal to at least one half of the face value of the defending force. This result demonstrates the common outcome of an attack: It succeeds, but with heavy loss to the attacker.

*Ex (Exchange).* The weaker force is destroyed, and the stronger force must lost participating units, the value of which is at least equal to the value of the weaker force. This result is very common and, obviously, penalizes the weaker force, whether it would be the attacker or defender. A variation on this has the stronger force taking losses only if it has a unit or units to lose that do not exceed the value of the defending units lost. If the only attacking units eligible for loss are stronger than the defending units, the attacker takes no losses in this type of exchange.

*Eg (Engaged).* The forces involved in this combat remain locked together, sometimes with further combat mandatory, etc. This is a more common outcome than most "gamers" would prefer, but is typical of many twentieth-century battles where two sides remain locked together in combat for long periods.

*Cn (Contact).* The forces involved have discovered each other, which means that in game systems using dummy or concealment counters, the two sides involved in the contact result are now revealed.

*Ca (Counterattack).* The defending force usually must make an immediate attack against the attacking force. This particular option is used to depict the lethality of a unit's situation once it has successfully penetrated into an enemy line.

*Pk (Panicked).* The affected force (can be either attacker or defender) is panicked and usually suffers a result that will hinder its effectiveness either by "freezing" it (cannot move or attack) or by forcing it to wander off when its turn to move next comes around. Commonly found in tactical-level games.

*Rt (Routed).* The affected forces execute a headlong retreat, a variation on the effects of panic, disruption, etc. A rout effect often has a similar panic effect on other friendly units encountered as the unit flees to the rear. Usually only found in tactical-level games.

*De (Defending Units Eliminated).* Identical to all of the results given to the attacker such as Ae, Ar, etc.

There are many more results and combinations of results possible in wargames. The ones listed above are the most commonly found, but even in that short list you can see how easily one may come up with variations.

• **Combat Supply.** A unit having a specific amount of supply enabling it to attack or defend normally. This is the ammunition and fuel-type supply for combat as opposed to food and other types of noncombat supply.

• **Combat Unit.** A class of game units distinguished by their ability to engage in combat (usually able to move and operate independently also). These are the "fighting" units as opposed to the "administrative" units.

• **Command Control.** A rule whereby units "in" command control may function normally while those "out" of command control have their ability to move and/or have combat impaired to various degrees. This represents a loss of the commander's ability to communicate with and control his units.

• **Command Radius.** The number of hexes from a leader unit that other units may be influenced by the leader unit.

• **Communications, Line of.** A line of connected hexes, free of enemy interference, that can be traced from the "supplied" unit to the source of supply. The supply line along which supplies are transported.

• **Consolidate.** Combining two or more smaller units into one larger one. This is often done in such a way that the strength of the larger unit is greater than the sum strength of the smaller units. This represents the greater organizational strength of the larger unit.

• **Continuous Line.** An unbroken line of friendly units (or Zones of Control). Often a necessary rule to represent the need for a "front line."

• **Continuous Turn.** A concept whereby one side can continue to move and fight until it runs out of success (or luck, usually both at the same time).

• **Controlled Hex.** A hex upon which a unit is exercising an effective Zone of Control. Usually any adjacent hex, but may also include hexes farther out, or not all adjacent hexes.

• **Counter.** Another term often used for the playing pieces in the game.

• **Covering Terrain.** Terrain that provides concealment from enemy observation and/or protection from enemy fire. Usually found in tactical-level games.

• **Dead Pile (or Dead Box).** Refers to the common practice of obtaining reinforcement or replacement from among a pile of units that have already been destroyed (the ''dead''). Macabre, but pragmatic.

• **Defender.** The player (the units belonging to that player) who is the object of an attack. Generally this is the nonphasing player (the player who is not moving).

• **Defending Strength.** The numerical rating of the ability of a unit to defend itself against enemy attack strength. Expressed in terms of defense-strength points.

• **Demoralization.** A condition encountered by an army (or individual units) when it has taken very heavy losses. The usual result is a loss of effectiveness that is expressed in terms of reduced ability to move or have combat. The exact details vary much with each game using such a rule. When a unit has reached the point where it is demoralized, it is said to have reached its demoralization level (''threshold''). For an army, this is measured in terms of strength points or units lost.

• **Depot Unit.** A specialized military unit (either mobile or stationary) that does little more than provide supply for other units. Loss of these units reduces the owning side's ability to move and fight.

• **Direct Fire.** Usually found in tactical-level games. The fire of weapons that have a flat trajectory. What they see is what they can try to hit.

• **Display.** A chart on which markers or playing pieces may be placed for the purpose of keeping track of information vital to the play of the game.

• **Disrupted.** In some games, there is a combat effect that, in effect, ''stuns'' the affected unit for one or more turns. The unit counter in question is either flipped over or has another marker placed on it to indicate the disrupted state. Disruption usually lowers movement and combat ability and sometimes the Zone of Control effect.

• **Diversionary Attack.** Usually found in games where all friendly units adjacent to enemy units must attack. The purpose of the ''diversionary'' attacks is to distract enemy strength so that the more important attacks have a better chance of succeeding. See entry for ''Soak-off.''

• **Dummy Unit.** A playing piece (usually with printing on both sides) that looks like a real unit on one side but that is really a deception that is only discovered when it is too late.

• **ECM (Electronic Countermeasures).** Using electronic devices to disrupt enemy electronics (radar, radio, and other sensors and communications devices).

• **Elevation.** In tactical-level games, the map often represents different heights (hills, buildings, etc.). There are often quite complex rules to deal with blocked lines of sight.

• **Enemy.** The opposing player and his units. Common term in game rules.

• **Engineer Unit.** The primary function of such units in games is to either construct fortifications or assist in their destruction or otherwise assist in getting through enemy defenses. Also used to construct things in general (bridges, bases, roads, etc).

• **Entry Cost.** The movement-point cost to enter a particular type of hex.

• **Exit Cost.** The movement-point cost to leave a particular type of hex. Either the cost is paid to enter a hex or leave it in a particular game.

• **Facing.** The direction a unit is facing, expressed in terms of which hex side a particular side of the playing piece is facing. Important in tactical level games where ''front'' and ''flank'' are vital.

• **Fire Strength.** The numerical rating of the missile-firing (a unit that can attack units that are not adjacent) ability of a unit. For example, guided-missile unit, longbow, etc.

• **First Player.** The player who takes action first in the sequence of play.

• **Flight Simulator.** A computer wargame from the perspective of the pilot of a combat aircraft. A joystick or a keyboard is used to control the aircraft as what a pilot can see from the cockpit appears on the screen. Vehicle and ship simulators are similar, except the player is inside a tank or ship.

• **Force.** A term used to refer to a group of friendly or enemy units.

• **Formations.** The manner in which units are laid out on the map. A column formation has all units advancing in a line one behind another. A line formation has all units lined up facing the enemy units. There are many other variations. Most often used in tactical games.

• **Fort (or Fortification).** A location printed on the map (or represented by a counter) that indicates a defensive emplacement that will increase the defensive power of the units using it.

• **Free Deployment.** An initial placement of units that allows the player to position the units any way he wishes (as opposed to strict historical placement of units).

• **Friction.** In some games, units are allowed to keep moving beyond their normal movement allowance. The catch is that when you push a unit like this, ''friction'' sets in. The effects of friction are similar to the damage caused by attrition, combat, and/or disruption. In actuality, friction is what happens normally when things begin to break down from use.

Driving a car for 100,000 miles causes a certain amount of "friction." If you drive the car 100,000 miles in one trip, you're going to have some more "friction." Same with military affairs, although more common. There is always some friction in military operations, and often there's a whole lot.

• **Game Assistance Program (GAP).** As PCs (and even programmable calculators) became available in the 1970s, wargamers developed programs that would automate the more cumbersome elements of manual wargames. GAPs never caught on as commercial products, but many were created and used.

• **Game Map.** The playing surface, usually covered by a hexagon (or other-type) grid to regulate movement and position of units. In most games, this is actually a map, similar in many respects to any other kind of map.

• **General Supply.** The type of supply that simply allows a unit to continue to exist. Often "attack supply" is required in addition to launch an attack. There are many variations on how supply is handled.

• **Grain.** The direction in which the straight rows of hexagons appear to run parallel to one another or the edges of the map. "Short grain" means that the grain runs across the narrow portion of the map. "Long grain" means it runs across the long portion of the map. Most maps are short grain because it is easier to attack "with the grain," and the defender in most games needs all the help he can get.

• **Grognard.** Not a technical term as such, but a term you'll hear in wargaming. It refers to experienced (and, these days, often middle-aged) wargamers. The term was originally used as a nickname for members of Napoleon's Old Guard. The term is French and means, literally, "grumbler." It reflects the attitude of the veteran troops who knew what was really going on but couldn't do much about it. So they grumbled, and so do most wargame grognards.

• **Ground Support.** The use of tactical aircraft (bombing, etc.) to assist friendly ground units. In most games, this is treated as just another form of artillery (indirect fire).

• **Headquarters.** Usually a unit that has no combat value itself but whose loss would decrease the capabilities of your combat units. These penalties usually take the form of an inability to move some combat units or to decrease their combat strength (especially when the "headquarters" unit actually represents a logistics [supply] unit).

• **Hex.** A single hexagon in a hexagonal grid.

• **Hex Number.** In many games, a four-digit number that describes the row the hex is in (first two digits) and the hex in that row (second two digits). It is printed in the hex. It was developed by Arnold Hendricks (while he was working for wargame publisher SPI in the early 1970s) and later released to the public domain for any game publisher who wanted to use it (many do). A variant of the older letter-number hex-identification system.

• **Hex Side.** One of the six sides of the hexagon. Sometimes rivers, ridge lines, fortifications, or some other feature run along the hex side. This means the gamer must pay attention to the differences in types of hex sides as well as different hexes.

• **Holding Area.** A box printed on the map for holding units that are not being used in the game (but will be, for one purpose or another).

• **Indirect Fire.** Fire of a unit that cannot be seen by the defender (a mortar, as opposed to a rifle).

• **Initiative.** As in real life, this is the ability of one side to act first. Usually, each side has a probability of moving first (or second, if that is preferable) and this probability often changes with success and failure on the battlefield. An initiative rule often gets quite complex, as one would expect from as complex an item as initiative.

• **Interdict.** Fire that interferes with enemy units' movement or supply, usually accomplished by air support or artillery indirect fire.

• **Isolation.** A unit is cut off from supply or command control. Usually, this occurs when enemy units of Zones of Control prevent a unit from tracing a line of hexes back to a supply or command source.

• **Leaders.** While wargames allow the player to take the place of the original commander, in some situations the actual location of the commander (or his subordinates) was vital, and this is represented by playing pieces representing the leaders. These "units" usually have no combat capability themselves, but rather possess the ability to increase the effectiveness of real combat units. Typical effects of leader units are to increase the movement or combat capability and to rally units that have been disrupted.

• **Line of Sight.** The path of unobstructed vision between two hexagons on a game map. Rules for this feature are often quite complex, but basically all you want to do is figure out whether any hex containing an obstacle (trees, buildings, etc.) is in the way.

• **Maneuvers.** In some games, there are certain prescribed sets of movements that a unit may or must perform in a certain way. For instance, in an air game, an aircraft might have to go from one altitude and head to another to perform a Split S maneuver, or a naval game might require that ships "follow the leader" when turning in line.

• **Mechanized Unit.** A military unit, usually a combat unit, which is completely motorized—that is, all men and equipment are in a vehicle of some sort (armored or unarmored).

• **Melee Combat.** Hand-to-hand combat, with the sword, spear, shield, knife.

• **Miniatures.** A form of wargaming (the earliest, in fact) where the playing pieces are lead figurines and the playing area a three-dimensional model of the battlefield. These games are almost always tactical level.

• **Morale.** This is the troops' state of mind, or how willing they are to push on in the face of the stresses and horrors of warfare. In games that use morale rules, combat and strenuous noncombat operations (forced marches, lack of supplies, etc.) will lower morale and, ultimately, the ability of units to function at peak level. Usually, a victory or rest or the presence of a powerful leader will restore morale.

• **Movement Allowance.** The numerical expression of a unit's ability to move, expressed in terms of movement points, with one movement point enabling a unit to enter a "clear terrain" hex on the average game map. Most other types of hexes usually require more than one movement point (or, sometimes, less).

• **Operations Points.** A variable number of "points," some of which are needed for any action in a game. Only a few games use operations points, and these are usually those that have continuous turns.

• **Opponent.** The other player or, as you get into the game, all of the forces of the "other side."

• **Opportunity Fire.** In tactical games, this is a unit that is ordered to fire only if enemy troops enter a certain area. Accurately represents a common technique in actual warfare.

• **Order of Appearance.** Lists when new units enter the game, and where.

• **Order of Battle.** The makeup of a player's forces in terms of types of units and the number of each type available.

• **Order of Battle Sheets.** A printed form with the game units displayed along with boxes to check off the declining strength of the units as they suffer combat (or noncombat) losses.

• **Organic.** In military usage, a weapon or unit that is normally assigned to a unit. In other words, a standard part of that unit.

• **Overrun.** An attack of such overwhelming force that the defending unit is destroyed so quickly that the attacking unit is able to continue movement with only minor interruption. An important rule in games in which each side must first move all its units and then attack. A recent example of this was seen in the 1991 Gulf War, where U.S. units advancing into Kuwait generally rolled over defending Iraqi units.

• **Owning Player.** The player to whom something (a unit, a section of the game map) belongs.

• **Party.** Not exactly a social event. This term refers to the group of characters that operate together in a role-playing game (RPG). Each character often has very different capabilities that, in theory, should complement each other.

• **Phasing Player.** The player whose phase it is at that particular point in the game. This is an important distinction because some sequences of play have the two (or more) players trading back and forth the ''active'' role.

• **Plot.** A written-down record of what one player's forces are going to do in a turn of a game using simultaneous movement. It's not as tedious as it sounds, as such games have fewer than a dozen units per side and simple movements. This approach is common in many computer wargames.

• **Point.** The basic unit of measure for describing the quantity represented in the movement allowance, combat strength, and any other value used in most games—for example: movement point, combat-strength point, etc.

• **Point and click (or drag).** Used in computer wargames when the mouse is used to activate (click) or move (drag) things on the screen. If you have a computer mouse, you know what this means.

• **Production Center.** A unit or installation printed on a map that is used to produce new units or supply, etc.

• **Protection Strength.** The defense strength terrain adds to any unit in a hex with that type of terrain. Also used for more complex defense-strength situations. For example, a tank would have one defense strength against cannon fire and a different protection strength against missiles.

• **Rail Capacity.** The capacity per game turn of the player's rail network to move units and/or provide supply by rail. Measured in rail-capacity points. Railroads have been the primary means of moving military supply (and often the units themselves) over land during the last 150 years. Railroads continue to be of vital military importance.

• **Random Events.** Many historical situations have a host of minor (and not so minor) tangential events that can influence the main operations to one degree or another. Sometimes these are political or economic events back home, or lethal diseases or severe weather encountered during campaigning. Sometimes these random events are included in a game simply to spice it up.

• **Range Allowance.** The number of hexes (in any direction) that an air unit (or the fire of an artillery or missile unit) may project its power.

• **Range Attenuation.** The effects of range on a missile weapon. Archers, artillery, and even aircraft are less lethal the farther away their target is. Rules for Range Attenuation take care of this.

• **Reinforcement.** A unit that does not start the game on the map but enters later. These units are kept track of with a reinforcement schedule and/or a reinforcement track (a display upon which the reinforcement units may be placed).

• **Replacements.** Similar to reinforcements, except these are not new units but are used to revive existing units; they are expressed in terms of replacement points. Each destroyed or depleted unit could be revived with a certain amount of replacement points. Historically, units suffer only 20–50 percent losses before losing their combat capability. Such units can be, and usually are, revived with replacements and time away from the battle for rest and training.

• **Retreat Priority.** A ranked listing of hex types or hex conditions into which a unit should retreat when called upon to do so as a result of combat.

• **Road Movement Rate.** Usually the number of road hexes a unit may move on for each single movement point expended. This is the (usually) sole exception to the standard of one movement point being expended to enter the easiest-to-enter terrain.

• **Role-playing Game (RPG).** A type of conflict simulation in which the player does not command two or more "units." Instead, the player controls only one "unit" or "character" that has more detailed characteristics than units in multi-unit games. The players often play their character against an assortment of enemies and obstacles controlled by a nonplaying "gamesmaster." The most popular role-playing games are based on fantasy backgrounds, although since RPGs were developed in the mid-1970s, there have been several published that deal with strictly military subjects.

• **Rolling Up a Character.** A term used in role-playing games, it refers to the process whereby a player, using dice and probability tables, determines the levels of his character's various powers and abilities.

• **Scale.** The size of each hexagon and the real time represented by each game turn. The size of the hexagon is measured from side to side and varies from a few meters upward to many light-years (in science-fiction games). Three different scales for games are used for land (and to a lesser extent naval and air) games:

**1. Tactical.** This is often referred to as the "tactical" level. This includes everything from individual men equaling one playing piece (and each turn representing seconds) up to groups of men numbering 30–50, hexes of a few hundred meters across and turns representing as much as an hour or so. This is the level where the player is controlling small ground units (platoons down to individual vehicles and troops), individual ships, or aircraft.

**2. Operational.** The next level covers everything up to the point where "tactics" (the precise maneuver of units) become less important than "strategy" (the allocation of resources on a broad scale). A player would control units representing battalions up to divisions, or groups of ships or aircraft.

**3. Strategic.** The "strategic" level generally involves hexes representing dozens or hundreds of kilometers and time scales representing weeks or months. At this scale, the player is controlling several armies and often entire nations (including their economies and politics).

• **Scenario.** A complete description of the event (battle) to be simulated in the game. Often, a number of scenarios are represented in a single game. Some are historical, allowing the player to deal with only portions of the battle or campaign, while others are "what if" or hypothetical situations. The scenario will detail the units to be used, where they are to be placed, or when they will arrive as reinforcements. Finally, the victory conditions and any scenario specific rules are given.

• **Sequence of Play.** One of the more important parts of the game. The Sequence of Play details (in strict list format) which player is to do what and when during each game turn. The game turn is the basic time unit in a game. The parts of a game turn have a strict hierarchy, which generally follows this order:

Game turn
Player turn
Phase
Segment
Step

In some games (usually strategic level), game turns are combined into repeatable sets to allow for functions that take place every so many turns. For example, in a game with monthly turns, every three months (or every year) certain production events may take place. This group of game turns would be called a game cycle. In some cases, both players perform a function during some part of the game turn. In this case, the activity would be called "joint" (as in joint player turn, etc.). In some games this whole procedure gets rather involved, which is why so much organization is needed.

• **Soak-off.** Also called a "diversionary attack." Making unfavorable attacks adjacent to an enemy on which unit you have concentrated maximum force for a critical attack. In order to attack the one enemy unit you really have to get, it is often necessary also to make attacks on enemy units adjacent to the one you are after. This is most common when the game stipulates that all adjacent enemy units must be attacked each turn. This is a wargaming technique with a close analogy in history, where it is known as the "secondary attack" and serves the same purpose of distracting the enemy from your more important, or "primary," attack.

• **Stack.** A group of friendly units (including, where appropriate, markers) placed in the same hex. The number that can occupy the same hex is limited by the game rules.

• **Stacking Points.** Another form of stacking, in which a numerical value (the stacking value) is used to determine how many units can be placed in a hex. Each side (or each hex) is then allowed so many stacking points per hex. In this way, units of differing size (some may be five to ten times larger than others) may be accurately represented.

• **Standard Game.** Most games have a standard game (scenario) that presents the basic historical situation. The units are set up as they originally began the battle or campaign, and none of the usual optional rules that are added later to allow players to explore the various historical possibilities in the battle are used. The standard game is sometimes called the basic or introductory game. Players who are more knowledgeable about the historical event being simulated will often skip the standard game and get right into the more advanced versions.

• **Strategic Movement.** In some games, units may move much more than they would be normally allowed to by restricting their movement to friendly-controlled hexes. This represents the ability of an army to use its transportation capability to its maximum potential, mainly because it doesn't have to be constantly ready to encounter and fight enemy units.

• **Strength Step.** In most games, an adverse combat result will either move the affected unit out of its position or destroy it. In some more complex games, instead of the unit being destroyed, it merely loses a portion of its strength. In a case such as this, a unit's strength is usually expressed in terms of strength steps. When the affected unit takes a step loss, a new counter with the same identification but lower combat strengths (and sometimes movement strengths) is substituted. When using such a rule, each unit has normally two, three, or more steps.

• **Supply.** These are the consumable items used by a unit during movement and combat (ammunition, food, fuel, etc.). A unit that has supply is said to be "in supply," while a unit that is not being supplied is "out of supply."

• **Supply Line.** This is another term for line of communications (see Communications, Line of).

• **Supply Unit.** This is a source of supply that can itself move. The primary function of such a unit is to provide supply for fighting units in the field. Units such as this can normally defend but not attack.

• **Target Acquisition.** This denotes that a target for long-range fire is in sight and capable of being fired upon.

• **Target Hex.** This is a term used in tactical games to designate the hex occupied by a unit that is going to be attacked, often by nonadjacent attacking units. The target hex may not be occupied, but the player may want to block enemy supply lines, create a temporary Zone of Control, or destroy a map feature (bridge, road, airfield, etc.).

• **Task.** One of a list of specific actions that a unit may perform during a game turn. Any action that is not on the list of permissible tasks may not be performed.

• **Task Allowance.** Certain tasks that a unit may perform during a game turn require different amounts of effort. Units are given a numerical rating expressed in task points. Each task is also given a task-point price that a unit must spend in order to execute it. For example, a unit may have a task-point allocation of 20 points. To move one hex may require only one task point. To entrench the unit where it is may require 10 task points, while attacking another unit may expend all 20 task points.

• **Terrain.** The game map may look like a normal map in many cases, but in all cases it is actually a collection of hexagon-shaped cells, each hexagon containing a precise type of terrain. Each hex has a dominant physical characteristic that has a precise effect: upon movement and/or combat of the units entering it. These kinds of terrain are more specific in tactical games and more general in strategic games. For example, in a tactical game you would have hexes representing woods, swamp, sand, or clear (open) areas. Operational-level games (10–50 kilometers per hex) would have rough terrain and forest and mixed (rough ground and forest) terrain. Strategic games (50-plus kilometers per hex) would have two or three kinds of rough terrain and impassable terrain, etc. Rivers and streams usually run along the hex sides to make it unambiguous as to which side of the river a unit is on.

• **Terrain Effects Chart.** This is an important element of any game, which shows the effect on movement and combat for units entering the different types of terrain in the game. Movement is usually affected by a variable number of movement points required to enter the different kinds of terrain. Combat is usually affected through increasing difficulty for the attacker attacking the defender in more favorable (for the defender) terrain. That is, the defender might be able to double his strength in certain types of terrain or even triple or quadruple it, or conversely, he might simply be able to shift on the Combat Results Table from a 3 to 1 to a 2 to 1.

• **Turn Radius.** The amount of space required to turn a ship or aircraft (and sometimes armored vehicle in detailed land games) around.

• **Unit.** A common term (along with the term ''counter'') for a playing piece that represents a military organization. Military units on the ground come in the following sizes:

*Single Soldier.*
*Fireteam.* Three to six men
*Squad.* Eight to 16 men (there are usually two fireteams per squad)
*Platoon.* Thirty to 60 men (three squads)
*Company.* Three or four platoons (100 to 300 men)
*Battalion.* Three or four companies (400 to 1,200 men)
*Regiment or Brigade.* Three or four battalions (1,200 to 5,000 men). The main difference between a regiment and a brigade is that a brigade is often a bit larger than a regiment and capable of operating independently, while a regiment is normally part of a division.
*Division.* Three or four regiments or brigades comprise a division (6,000 to 20,000 men).
*Corps.* Two to four or more divisions are a corps of 20,000 to 70,000 men.
*Army.* Two or more corps constitute an army (50,000 to 250,000 men).
*Army Group.* Two or more armies are an army group.

The aforementioned types of units have been in common usage since about 1800. Before that, the terms "battalion," "brigade," and "army" were most prevalent. The term "battalion" came out of the term "battle," which was a group (varying from a few hundred to a few thousand) that could be controlled by one man and long called a "battle." In pre-gunpowder days, the Romans had a very flexible unit organization in which units they called "cohorts" were very much the equivalent of modern-day battalions. A group of cohorts made up a legion, which was remarkably similar to a modern-day division. All of the individual military organizations will show up at one time or another in a game. The non-land environment unit counters will either represent individual planes or ships or small groups of same.

• **Unit Designation.** The identity of the unit, usually a historical designation. An important part of the historical component of a wargame.
• **Victory Conditions.** Victory is often a fairly vague thing in military history, and one of the more important parts of a game is specific victory conditions to determine who has won. These usually consist of specific values to such things as the destruction of enemy units, the possession of terrain features, etc.
• **Victory Level.** Even when one has achieved victory, all victories are not equal. Certain levels of victory have been established through the

years. These levels are, in ascending order, draw, marginal victory, substantial victory, decisive victory, overwhelming victory.

• **Weather.** Often depicted in games, particularly with regard to its effect on movement. Games that cover more than one season in a region with severe winters often have to show the effects of weather. Tactical-level games often show the effects of fog and rain (limits visibility) and mud (limits mobility).

**[9.0] ZONES OF CONTROL**

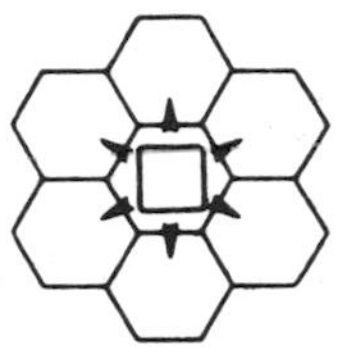

GENERAL RULE:

The six hexagons immediately surrounding a hex constitute the Zone of Control of any units in that hex. Hexes upon which a unit exerts a Zone of Control are called "controlled" hexes, and inhibit the movement of Enemy units. All units exert a Zone of Control (except as noted in the Cases below).

• **Zone of Control (ZOC).** Another one of the basic rules in gaming. Zone of Control represents the six hexes surrounding a unit that the unit controls. This allows some units to spread out as they would to cover more territory while concentrating others for a crucial attack. Zones of Control represent everything from the physical presence of parts of the unit in that hex to the ability of the unit itself to cover those controlled hexes with fire or to shift its weight in that direction should an enemy unit approach. Zones of Control may have many different effects on movement and combat to reflect the variables in the Zone of Control's usage for a particular game. The most common effects on movement of a Zone of Control are:

*Blocking.* Units must stop immediately upon entering an enemy-controlled hex and may leave only as a result of combat (either the enemy unit is destroyed or the friendly unit is attacked and forced back).

*Rigid (often called "Locking" ZOC).* Units must stop upon entering an enemy-controlled hex and may leave only at the beginning of a movement phase (usually it is not permissible to move directly from one enemy Zone of Control to another).

*Elastic (often called "Fluid" ZOC).* Units may enter and leave Zones of Control by paying movement-point costs just as they would for entering different kinds of terrain.

*Open.* Zones of Control have no effect on movement.

The various effects of Zones of Control on combat are as follows:

*Active*— This requires every unit in a Zone of Control to attack enemy units adjacent to them during a combat phase.

*Inactive*— Units do not have to attack.

There are also effects upon supply and the ability to retreat as a result of combat.

An *Interdicting Zone of Control* prohibits the line of hexes for retreat or supply from being traced through all enemy-controlled hexes, even if a friendly unit is occupying that hex.

A *Suppressive Zone of Control* prohibits the passage of supply or retreating units through a Zone of Control hex unless that hex is occupied by a friendly unit.

A *Permissive Zone of Control* does not affect the path of supply or retreat in any way.

For example, a blocking, active, interdicting Zone of Control is the most restrictive kind. Units must stop upon entering, may not leave except as a result of combat, and must attack any enemy units that are in their Zone of Control. In addition, units may not retreat into one of these hexes if forced to as a result of combat and may not trace any supply through them. On the other hand, a unit with an open, inactive, permissive Zone of Control in effect has no Zone of Control.

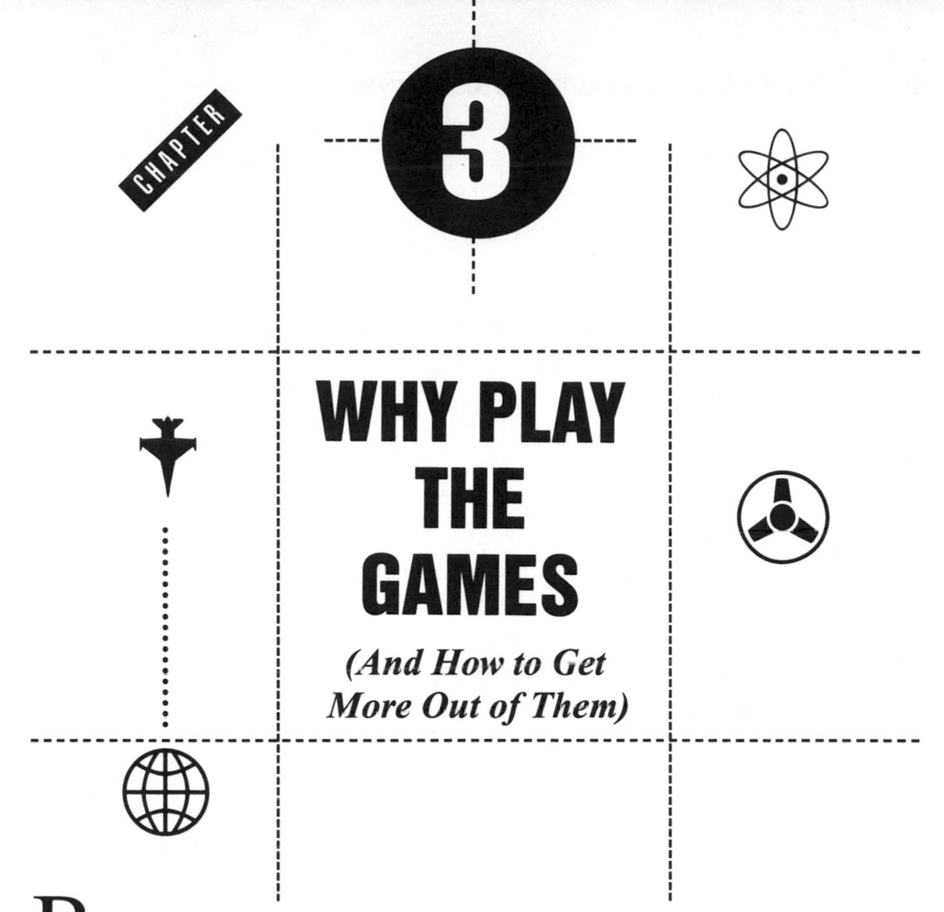


# WHY PLAY THE GAMES

## *(And How to Get More Out of Them)*

By far the most common reason gamers give for playing the games is to experience history. Actually, since simulation games also include many nonhistorical subjects (fantasy and science fiction, etc.) we might as well face up to the fact that experience of any sort is one of the most important things a simulation game has to offer. This experience consists of the gamer being able to massage information in order to see what different shapes the information is capable of taking.

The essence of a simulation game is that it allows, within well-defined limits, a great deal of variety in an otherwise strictly predetermined historical event. This is the popular "what if?" element in the games. For example, take General Custer's last battle at the Little Big Horn in 1876. What if he had, at the last minute, taken along his Gatling guns (primitive machine guns) after all? He could have taken the Gatling guns; thus, this is a reasonable "what if." He couldn't have taken any flamethrowers simply because he didn't have them. Note that what makes a fantasy game a fantasy game is that it is a game in which General Custer *does* go to the Little Big Horn with flamethrowers, and maybe even a death-ray gun.

Whether historical or nonhistorical, almost all simulation games contain four general kinds of information: geographical, Order of Battle, situational, and dynamic potential.

## INFORMATION ✱✱

The geographical information is usually obtained from a playing map, assuming that the game has one, and most do. Even those that don't have one often require that the player construct some type of playing surface. The Order of Battle is nothing more than a list of the units or entities that took part in the original event. The situational information is the scenario—that is, the set-up of the forces, the conditions of victory, and any additional factors such as the state of the weather, etc. The dynamic-potential information in a game represents the players' perception of the first three information elements in light of where they may go. The first thing a player generally does when he obtains any game is to lay it out and quickly digest geographic, Order of Battle, and situational information. Once the game is given a good looking-at, the player then obtains what he is really looking for, the dynamic potential of the game: where the game might go, and what it might do. Computers have yet another advantage in this area in that a player can even more quickly take a look around, although computer wargames also hide a lot from the player.

There are certain things the gamer should be aware of when he attempts to obtain information from a game. We tend to look for more in a game than is there. This is rather typical of most people's relationship with printed matter. The attitude is, "If it's in print, it must be right." Computers compound this because you can't even *read* the "print": The computer just does some mumbo jumbo inside the box and gives you an answer based on who knows what. Without casting too many aspersions on my work or the work of others, I would like to offer the following cautionary advice when you are examining games:

*Geography,* as with most things in wargaming, is very much subject to individual interpretation. There are two primary things to keep in mind when examining a geographical game map. First, it often has a grid, most often a hexagonal grid, superimposed over it. This means that each "cell" of geographical information is at least the size of a 16mm (diameter) hexagon. The normal "cell" of information in a traditional nongame map is normally smaller than a tenth of a millimeter (if you look closely at a map under a magnifying glass, you will see the little colored dots that the printing process uses to show where the water stops and the land begins). The second point is that in most historical situations, only very large ("gross") terrain features had any significant effect on operations. Thus, a great deal of detail on a map will often get in the way of providing an accurate simulation. The designer usually feels obliged to justify all of this detail. Often the gamer will be equally expectant that all of this detail be put to some use, or else why bother him with it? There is an unspoken assumption that only essential geography ought to be displayed. It is considered a bad design if information is included in the game that does not contribute to one's understanding of what is going on.

Given these two points, it should in most cases be fairly easy for the game designer to represent the key geographical elements of any conflict situation. The gamer, if he is sufficiently well read in the subject of the game, should expect that the geographical elements spotlighted in the written histories will also be represented in the game. If anything, the game will, in addition to confirming the importance of a particular piece of terrain, also show in more detail *why* the terrain was important. It is often the case that the game will show another element of the terrain that could have been important, but because it was not used, was not important in the original battle. An example of this is the Ardennes region of Belgium in 1940 and 1944. In 1940 the Germans dashed through the Ardennes with motorized forces, crossed the Meuse River, and proceeded to cut off the bulk of the British and French armies in Belgium. In this case, the Ardennes was not an obstacle, although it certainly could have been. This was proven in 1944, when the Germans again attacked in the Ardennes; this time they were not able to move right through, because American forces were already there and fought tenaciously, even though outnumbered. The Americans eventually stopped the Germans cold. Any historical simulation on the 1940 campaign would have to allow for the realistic possibility of the Germans being stopped in the Ardennes and then suffering the disadvantages of having to attack through that particularly rugged terrain.

Most designers, when they are dealing with terrain, go to one or more maps and do little more than place a blank hex grid over the map, shine some light behind the map and the hex sheet, and then trace the terrain features, making decisions on the spot as to what to include and what not to include. The only other decision is to make sure that the terrain is unambiguously distributed in the different hexagonal "cells." These terrain-analysis decisions are made on the basis of historical experience. It takes a little practice, although looking at existing games and historical maps of the areas the game represents will help.

The game designer then plays the game with this hex grid map. If it feels right and he does it right, that's your map. Often, additional detail will be added purely for historical interest. These details may be things such as additional town names that were mentioned in most historical accounts or the names of other geographical features that either do not necessarily need names (the names of forests or mountains) or might be useful for the player as points of reference when playing the game.

Computer-game designers generally use a scanner to copy the details of a paper map into a computer file that can then be edited to show what the designer wants to pop up onto the screen. Before scanner technology matured in the late 1980s, the programmer would painstakingly build a computer map with a "paint" program. The source would be a paper map, or even a wargame map on the same subject. Many computer-game maps use hexes to regulate movement and position (just like paper games). If the hex grid is not shown in the screen, it is implied. Some games give you the option of showing the grid or not.

The thing to keep in mind is that the game designer is not God. The game designer does not create terrain. He is merely supposed to represent it on paper. If he does it right, it will seem right to the player. If he does it wrong, you will sense this when you look at the map. To give you a better idea of the various ways terrain can be represented on a map, take a look at the Metz game map included in this book.

*Order of Battle* information can also do with a bit of scrutiny on the part of the gamer. As with the game map, the Order of Battle information must be "aggregated" a bit. As with the map, the scale of the game determines how much detail will go into the Order of Battle. The larger the scale, the more similar all of the units become. On the lowest scales (tactical), a much larger amount of differentiation among units is usually necessary. No matter what the level, however, not every specific unit that was historically present in a conflict will be likely to show up in the game as an individual unit. This is usually done partially for reasons of realism and partially for reasons of playability. Many small or otherwise functionally insignificant units are not necessary to the play of the game, and thus the designer will not include them. If there are a number of such minor units included in the game, I would consider that to be sloppy design since they will get in the way of play, depriving the player of the ability to learn from the game. A great number of insignificant units have a direct effect upon playability, and it is often a sign of good design when the number of playing pieces is kept to the minimum necessary to accurately represent the situation without unduly compromising historical information. Often, an effective compromise is to print a detailed Order of Battle in comparison to the actual Order of Battle used in the game.

Computer wargames have more options in this area. If the game has good Artificial Intelligence (AI), you can have a lot of small, but historically accurate, units running around. Good AI will relieve the player of the burden of telling each unit exactly what to do at all times. For example, good AI would enable you to order a dozen or so related units (say, all the battalions of an infantry division) to attack in a certain direction until an objective is reached or a certain level of losses is taken. The AI would then apply tactics appropriate for the troops you are commanding and carry out the orders. You could then go on to the next group of battalions or attend to some more critical aspect of the game. Not all computer games have this feature, and a major drawback of many computer games is trying to force the player to spend a lot of time dealing with many individual units.

The Order of Battle included with the *Drive on Metz* game is pretty representative of what you'll find in most games.

*Situational realism* is pretty straightforward. Once the game provides the geography in the form of a playing map and the units involved in the form of playing pieces, any player can (and some do) create the situation (game) from the written sources. In other words, at this point the player could just open a book on the battle and use the game map and playing pieces without even looking at the game rules. Many players do this. Computer wargames have had to cope with this situation also (aside from the problem of computer users not wanting to read instruction manuals).

Many games, especially the smaller ones, have but one scenario, and that is represented by the instructions for the game's one-and-only set-up. Any "what if" activity on the part of the player is expected to come out of the actual play. As with any historical game, the players may start out as the original participants did, but rarely will they make the same moves as the historical antagonists, and equally rarely will they come up with the same precise historical outcome (although often the side that won historically will win in the game). A more diligent designer will provide additional scenarios, each of which will expand upon some aspect of the situation that could have easily been different. Often, there were additional units for either side that could have been present at the battle or units that were present that might not have been there or might have been present in different strength for the historical battle. In addition, units may have been deployed differently. All of these elements should be included in a scenario. The scenario section is also a good place to display a lot of purely historical information. See the *Drive on Metz* rules for a sample of what a scenario looks like.

The *dynamic potential* of the game is the sum total of the players' overall perception of the game, particularly after examining its main elements. The rules are often fairly standard. This rules standardization is a common convention among designers, primarily to make it relatively easy for the gamer to play (and the designer to design). Each game does require a certain number of special rules that will cause the player to expend a certain amount of energy to absorb. Computer wargames follow the same pattern, with similar sets of commands and menus used in a great many games (often from the same publisher, or designer).

"Reading" games rather than playing them is quite common and always has been. Many gamers "collect" games. They buy them, but never play them. This does not mean that they are not used. Quite often, the hobbyist will spend several hours with the game. The usual procedure is to lay out the map, examine the pieces, read the rules and scenarios and perhaps place the pieces on the map, but that is generally as far as it goes. The player has been satisfied with experiencing the dynamic potential. By dynamic, I mean how and to what limits the various elements of the game may be manipulated. Every historical situation has this dynamic potential. Most history books or films are presented as a linear rendering of what went on, so there is no potential for exercising this dynamic. A game, of course, is just the opposite. Its elements are meant to be exercised, and a player will often do this in his head with the aid of the game components.

In terms of player interaction, there are three types of games: the one- or two-player game (by far the most common), the multi-player game, and the role-playing game. Each of these has a different dynamic potential. What I spoke about above was primarily the one- or two-player game. The multi-player game has a markedly different dynamic potential in that you are not only playing against a historical situation, but also against a group of individuals. This group (three or more people) creates a dramatically different dynamism in the game as the individuals are less restricted by the game situation because so much of the activity is dependent not on the game system but the cooperation of one or more of the other players. For this reason, multi-player games tend to be essentially simpler in their mechanics than the traditional two-player game. There are very practical reasons for this.

Since more people are playing, the sequence of playing for any individual causes every one of the other players to wait until that sequence is finished. If the total time it takes for every player to get through his sequence of play is too long, most players will lose interest. This is not to say that realistic multi-player games are not available. It is just that the dynamic of the game is against it. Thus, the dynamic of an individual gamer examining a multi-player game is normally considerably less than that of a two-player game. This is not the case with multi-player games played on remote computer systems like GEnie, private BBSs (bulletin-board systems), Prodigy, or CompuServe. As these games are computerized and player actions (nearly) simulations, they are quite lively and addictive. The downside is the price: six dollars an hour and up (except for some games on Prodigy, which has a low flat monthly fee, but has a dismal selection of offerings).

Role-playing games, on the other hand, have considerable dynamic potential when viewed by an individual player. This is one reason for such games' popularity. Most role-playing games to date, however, have been almost exclusively fantasy or science fiction in nature. This is mainly because the first successful role-playing game published (*Dungeons and Dragons*) just happened to be based on a fantasy-type situation (mostly the medieval-influenced Middle Earth world created by Tolkien). The essence of the role-playing game is that one player, the gamesmaster, does most of the work. The other players have a relatively simple "menu" of activities in which they may partake. These activities are generally restricted to movement and responding to whatever the gamesmaster throws at them. The gamers' responses take them into rather extensive rule booklets that consist primarily of charts and tables crammed with information.

An appropriate analogy would be a class in auto mechanics in which the instructor, using a demonstration engine, continually made certain components of the engine malfunction and then challenged the students to go through their manuals as well as their previous experience to come up with the most efficient solution to the problem he had created. Historical role-playing games have been tried. In 1979, *Commando* and *NATO Division Commander* were published. A few others followed in the 1980s. The RPG concept applied to wargaming subjects never really caught on. There are some military-oriented computer RPGs, with more in the works all the time. But, again, fantasy beats reality in the marketplace even when a computer is used.

## HISTORY ✱✱

There are other aspects of a game that provide opportunities to examine this kind of information from different angles. One of the most important of these is the historical period that a particular conflict occurs in. Specific key elements vary with the period: Ancient warfare had different key elements from Napoleonic wars or World War II. What follows is a discussion of all of these specific periods and areas and what the player should look for in each of them.

### THE ANCIENT PERIOD (3100 B.C. to A.D. 600)

It is unfortunate that the market is so thin for games in this period that we must cover such a sweeping era in human history in one "period." Ironically, demand for games of this period demonstrated a dramatic increase in the wake of Communism's collapse in 1989. With the resulting demise of popularity for all those "Warsaw Pact/NATO" games, many gamers suddenly became very eager to know more about ancient history. Very curious, but a true fact (based on consumer surveys and game sales).

While there was a lot of activity in this period from the viewpoint of a student of history, most of the military developments in this period were basic in the extreme. This era starts around 3100 B.C., with the beginning of recorded history. Much of our information of the earliest period is centered around the civilizations of the Middle East, primarily Egypt, and much of it's derived not directly from written sources but from archaeological findings.

During this time, the organization of armies had not progressed very far. When it came to operations on a strategic level, however, large kingdoms and empires were being formed and maintained over generations. Thus, really large armies, with troops numbering in the thousands, could only be raised and maintained for as long as they could feed themselves in the field. For this reason, most of the campaigns in this period were little more than large-scale raids. The objects of these foraging expeditions were somebody else's food, utensils, treasure, women, slaves, or what have you. War in this period, and indeed up until quite recently, had to pay for itself. Any games treating warfare in this age have to deal with the fact that the armies could literally die because of errors in the leaders' calculating where the food was or wasn't.

Warfare in this period was also very much an extension of local politics. These raids, which passed for campaigns, were conducted for the furtherance of political goals: to punish a neighbor, weaken a neighbor, steal from a neighbor, frighten a neighbor, assist an ally, force someone to become an ally. The threat of a few thousand hungry spearmen descending upon one's farmlands was usually a strong motivator for a change in political outlook. The kings of Egypt were never reluctant to use such a tool, and because of the fertility of their farmlands along the Nile, it was quite possible to go to war any number of times during the year, as long as this occurred between the planting and the harvest. At other times, the king had only a small bodyguard at his disposal, and it would have been foolish in the extreme to risk his person in a foreign land with such a small armed force. Professional armies were not going to appear for a while yet.

On a tactical level, things weren't much more encouraging. Chess, I should point out, is a rather accurate rendering of tactical warfare in this period. Indeed, it is a fairly accurate rendering of warfare (with some exceptions, which we will get to) up until the last few hundred years. To see the chess connection, consider what two armies had to go through in order to fight a battle in this period. After the general had managed the miracle of getting his army to meet at a certain place (where the enemy also happened to be) and the other side had deemed it a good idea to fight a battle, both armies would line up facing one another, normally with their mass of spearmen in the front ranks, archers, if any, just behind them. Cavalry would be on the flanks, with perhaps a reserve of the most trustworthy spearmen behind the archers under the direct command of the king or general himself.

It's important to note that, throughout this entire period, very few battles could be fought except by mutual consent. The amount of time required to get the troops lined up in any sort of battle formation might be five or six hours or more. During that time, the other army could simply march off (and often did, if the prospects did not appear promising).

Cavalry was not much superior to infantry in this period. Some combat was possible from the back of a horse but not much, and the foot soldier was not at the same disadvantage he would be after the development of the stirrup and the more heavily armored horseman. The stirrup was invented about 2,000 years ago, but up until about A.D. 500 most cavalry did not have it, which meant that the horseman was rather insecure on his mount. Therefore, it was possible to use cavalry for little besides rapid transportation for what was essentially a foot soldier. Horses tired easily, and had to be fed and watered besides. Large all-cavalry armies did exist in this period, but only in large grassland areas. Even then, these masses of nomads could only operate during warm weather, when there was plenty of grass to feed the horses. Since most of the early civilizations formed around heavily farmed river deltas, there was not a lot of well-watered grassland nearby to support large numbers of mounted nomads. Any nearby nomads were few in number compared to the masses of farmers and city dwellers.

The battles of the period depended somewhat equally upon the stamina and resolve of the opposing forces and the charisma and skill of the opposing leaders. The leader generally couldn't move too many units at the same time (no radios) and usually could move them only in a forward direction. A more skillful army would be noted for its relative abundance of effective leaders. But no matter how many effective leaders there might be in the entire army, communications was always an enormous problem. The battle line might extend for a few hundred meters with the troops shoulder to shoulder and anywhere from half a dozen to a couple of dozen ranks deep. An army containing 3,000 spearmen, 1,000 archers, and 1,000 cavalrymen (or chariots) would be a fairly substantial army indeed. Fully deployed, it might extend for only 200 or 300 meters plus cavalry or chariots on the flank and a small reserve (often the royal bodyguard) behind the battle line with the king. A more highly organized army might have its spearmen divided into contingents, generally under the command of the noble who raised them. Feudalism is an ancient institution.

With all the noise of 4,000 or 5,000 men (plus horses) standing about in such a small area (even if they were under orders to keep the noise down), sending a voice signal more than 50 or 100 meters would be chancy at best. Messengers could be used and often were, but they were relatively slow. Even more effective was the use of musical instruments and flags. It was not just for show that ancient armies marched off to war with a lot of flags and other pole-mounted signaling devices as well as a fairly substantial corps of buglers, drummers, and other musicians.

But even with this type of signal corps, the range of orders that could be issued was rather limited. Generally, one could order advance or retreat. A retreat was always a rather risky operation, as it might easily turn into a rout. Other key orders would be the commitment of the mounted troops or the royal bodyguard (often with the "royal body" as its head, and this was rarely done for obvious reasons). Once everyone got his troops sorted out and headed in the right direction, the fighting itself was rather desultory. In fact, the casualties tended to be rather light during the combat phase of the battle. Much depended on morale.

During this era, the bulk of military history shows that the defense had an advantage over the offense. Even the poorest spearmen generally had fairly large shields that they carried with them. If they could afford it at all, troops would have some rudimentary "armor" constructed of wood, multiple layers of cloth, wicker, or if they were really wealthy, leather. Metal, being such an expensive commodity, was usually used only by the wealthiest. Some type of protective head covering was also employed. The offensive weapons, at least up until about 2000 B.C., consisted of spears and daggers. Up to that point, metalworking had not yet reached the point where swords could be forged. Archery was also employed, but the bows were weak, the arrows crude, the arrowheads themselves being only mild bronze. For the most part, however, archery was employed in massive fire straight up into the air and down onto the enemy troops. If the defenders had their wits about them, they would raise their shields and give themselves fairly effective protection. The front ranks would have their shields lowered to face the enemy spearmen. Once the two masses of spearmen met, there would be a lot of jockeying about, pushing and shoving, and a lot of other activity, the chief product of which tended to be fatigue and demoralization. It was simply a question of which side broke first.

I suspect that the side that could put on the most fearsome show (including the rival musicians, flag bearers, and someone who could best be described as a "stage director") would prevail until the other side decided it wasn't worth it and tried to head for home simultaneously.

Most casualties were inflicted when one side began to hustle for the rear. It's one thing to try to get a shot at a fellow who's got a big shield in front of him. But when his back is shown, wound-inflicting prospects improve immensely. As one would imagine, the fleeing enemy has more of an incentive to move quickly than does the pursuer. The army in flight consisted of many individuals in danger of being either killed or captured and sold into slavery (or worse, as torture and buggery were popular postbattle activities).

It was during the pursuit of a broken enemy that the cavalry or charioteers came into their own. Chariots were expensive propositions: two-wheel carts drawn by two horses or other fairly swift draft animals (asses, donkeys, etc.). These vehicles were at their best in the pursuit of a broken enemy or in fleeing from a triumphant enemy. Then as now, the lowly G.I. did most of the work and constituted most of the casualties. The nobility and better-off soldiery tended to be mounted and made a major contribution only when the foot troops had won the victory. If the foot troops lost, the wealthier and more mobile soldiery would flee the battlefield largely intact, while their less noble companions were slaughtered by the enemy charioteers and cavalry. Now you can understand why being in the cavalry was considered such a swell situation.

Throughout this period, the most common form of combat was not the battle between two armies on a dusty field, but the siege. It was not for nothing that the first cities were noted for their walls. In fact, the walls came first, as without the walls anything within the city limits was easily seized by a superior army. Typically, an army would not invade a neighbor unless pretty sure that the victim could not field a larger force. The usual routine was to march in, grab anything worth stealing (food, animals, slaves), and march up to the city. If enough food was on hand, the invading army would lay siege to the city. The only thing that would prompt an assault on the walls was the assumption that there were too few defenders to resist and/or a shortage of food among the besiegers. A well-defended city maintained food and water reserves for a siege. If these were adequate, and if there was sufficient warning of the approaching army, the local farmers would flee with animals and additional food supplies to the city (burning any unharvested crops as they did so). Obviously, if an invader was quick enough, he would arrive at an ill-prepared city, short of food and local manpower needed to defend the walls. Sieges were risky, but skilled generals were able to bring them off. The price of failure was high, as a city not taken might send its own army to your walls in the near future.

Between 1000 and 500 B.C., a number of military "revolutions" took place. Around 1000 B.C., the age of iron really got rolling. Iron allowed for a number of improvements in weaponry and equipment. The introduction of iron swords made it possible for a more skillful infantry army to reach a decision without the pushing and shoving match that usually occurred in previous battles. The Assyrians came along with a number of innovations during this period, one of the key ones being archers on horseback. This opened up all manner of tactical possibilities, the principal one being highly mobile, long range firepower.

The Assyrian kings also instituted a number of other innovations, which were not so obvious on the battlefield, except for the results. First of all, they saw to it that their troops were always well equipped and that weapons were kept in repair. This was not always done in the past, and when it was found out, it was often too late. The Assyrians also instituted an ongoing training program for their troops (who were still, as were most armies, basically a part-time militia.) Superior organization also allowed the Assyrians to put more men into the field for longer periods of time. Although accurate numbers are impossible to come by, a good estimate is that the Assyrians were capable of fielding armies of upward of 30,000 to 40,000 men.

Another interesting innovation of the Assyrians was the use of mass terror. They would literally kill everything in sight if there was any resistance when they invaded an area. The shape of things to come, one might say. The word quickly got around, and before long all it took was the approach of an Assyrian army to add another victory to the Assyrians' win column. They eventually ran out of local enemies, got soft, and their former victims eventually evened the score. There aren't many Assyrians left. Those that do remain live in Iraq, where they are no longer noted for their military prowess.

Meanwhile, back in Greece and off in Rome, many of the same basic ideas the Assyrians had developed were being reinvented. By 500 B.C., warfare in this region was becoming very highly organized. In fact, at this point warfare achieved a degree of organization that was superior to what would exist a thousand years in the future.

The armies of Greece and Rome (not to mention Persia) were models of modern organization. The organizations were rather formal (with units of 100 men, 500 men, 2,000, 3,000 men, etc., depending upon the army), with groups of leaders functioning as the trained officers for (and especially among the Romans) the beginning of a regular, fairly large standing army. At the height of the Roman Empire, between A.D. 100 and 200, the Romans maintained regular forces of more than 100,000 men. This, in an empire containing perhaps 100 million citizens, was an unheard-of feat. It was for this reason, in addition to their overall administrative skills, that the Roman Empire lasted as long as it did.

Despite the increasing efficiency of field armies, the defense managed to stay ahead of the offense. This was primarily through the more thorough development of the art of fortification. Fortified cities and military camps became extremely difficult, and expensive, to take. It was not impossible, though, and the professional armies that were becoming more prevalent (the Greeks of Alexander and the Romans of the empire period) generally had no trouble taking any fortified place if they wanted to badly enough and were willing to pay the price.

The most effective of the ancient armies still had to face the same basic problems of any ancient army: poor communications (over long distances), mobility limited basically by how fast the men could walk (or travel by ship), and complete dependence on the local food supplies. These were harsh restrictions that were not really overcome until the last century.

The superiority of the Romans, for example (as well as most other successful ancient military organizations), was mainly a result of their more effective killing methods. The basic Roman fighting unit, the cohort (about 500 men), developed a battle drill in which the first two ranks of troops would vigorously engage the enemy with their swords and shields while the rear four ranks would throw their spears and hold themselves ready to relieve the two ranks fighting. This constant "assembly line" approach to hand-to-hand combat was usually invincible against less well trained opponents. Indeed, it was not superior technology and skill that overcame the Romans eventually. It was sheer force of numbers. Even though the Romans tried (often successfully) to absorb the barbarians surrounding the empire, a combination of internal decay and tremendous external pressures brought it down. With this event, the ancient period came to a close. All was lost, and the Dark Ages ensued.

The main things to look for in the ancient period are games that emphasize the politics and economics of warfare. These subjects were complex and interesting even way back then. For games on battles, look for an understanding of the details of these actions. Many of the military systems were not as simple as they looked.

### THE DARK AGES (600 TO 1200, ALSO KNOWN AS THE MEDIEVAL PERIOD)

The Dark Ages were overall a step backward in terms of military technique and technology. The one big innovation was the increased prominence of the mounted fighting man. From 100 to 500, there was a seemingly never-ending wave of nomad horsemen armed with swords, spears, and bows coming out of the central Asian plains. The horse archers would fire volley after volley into the foot troops. Then, when the defenders seemed suitably weakened, mounted lancers would charge in. With the aid of the stirrup, the shock effect of these horse lancers was nearly irresistible. Non-Oriental barbarians in Europe, particularly the Germans, adopted these Oriental techniques, and out of this came not only the destruction of the western Roman Empire, but also the development of a mounted, armored "man-at-arms." This included the legendary "knight in shining armor," but most of these guys were simply well-trained and experienced swords for hire.

Because of the tremendous breakdown of organized society in Western Europe, the European mounted warriors tended to be fairly independent individuals. They leaned toward more armor and less archery, not to mention less organization and, in large groups, were less effective. But since men-at-arms generally fought one another it really didn't matter most of the time. Gone was the ancient pattern of warfare whereby the vitally protected infantry would do most of the work and take most of the casualties while their better-armed armored (and mounted) "betters" occasionally hacked away at one another, but more frequently pursued the enemy or left the battlefield when danger threatened. The medieval man-at-arms mainly wanted to go after another man-at-arms. Any lowlife infantry in the way were usually just ridden over (even if they were friendly).

Such superior military organizations that existed during this era generally followed the Oriental (Mongol or Byzantine) model. That is, they used fairly disciplined forces consisting of mounted or unmounted archers and equally disciplined and armored fighting troops armed with lances, shields, and swords (or axes, or whatever).

In a time when there was a lot less wealth to go around, campaigns and battles became less frequent. The Byzantine Empire lasted until the 1400s primarily because it had one of the more efficient fighting forces. But the mainstay of the Byzantine Army was the cataphract. This was a mounted and well-armored fighting man equipped with a very powerful bow, a lance, a heavy sword, a shield (with three heavy and fairly deadly darts behind it), and sometimes even an ax. The cataphract could fight mounted or dismounted. He could use his bow either way also. The cataphract was a pretty deadly combination and probably the most effective of the medieval warriors.

The battles themselves during this period hadn't changed much from the earlier period. There was still the communications problem, the battle still had to be by mutual consent (except when one side was mounted and the other wasn't), and most armies still depended to a large extent on part-time warriors, although the level of skill of some of these part-time armies was quite high. Since fighting seemed to be the thing a lot of these warriors would prefer to do, other occupations were dull but necessary.

This epoch was noted for some rather dynamic operations. The barbarian nomads continued to come in from the East, and every few generations there generally would be some sort of campaign and battle. The battles tended to be quite dramatic. The campaigns could be rather dull. Within this era of 600 or 800 years, there were still plenty of things that a gamer will find interesting, particularly the economics and politics.

### THE RENAISSANCE PERIOD (1200 TO 1600)

As far as military history goes, this era was not so much one of renaissance as it was of transition from primitive warfare to the rudiments of modern warfare. Most of the military institutions, tactics, techniques, and weaponry that we recognize today had their origins in this period.

Infantry, for example, regained its premedieval primacy. It did so, ironically, by going back to the weapons, tactics, and formations of ancient Greece—that is, the phalanx. This was a formation of spearmen. As with the Greeks, some of the spears were as long as 21 feet; there were a dozen or more ranks and every rank thrust its spears forward. The "spear wall" presented an unbreachable obstacle even to the heavily armored cavalry. Although advances in metalworking and armor design made the armor protection of the "men-at-arms" nearly invulnerable, their lack of training and discipline did them in as much as anything else. The pikemen of the Swiss and other German states were, in contrast, highly disciplined and organized. It was this, as much as their use of the spear wall, that enabled them to revolutionize ground warfare.

The next big innovation was to be found in the proliferation of missile weapons. The longbow of the English was only a temporary advantage as it required a considerable amount of training and practice, and this severely limited the number of longbowmen who could be fielded. The crossbow, although not nearly as effective in terms of rate of fire and range as the longbow, was able to penetrate armor and could be effectively used by relatively untrained soldiers. For this reason, the crossbow probably saw far greater use than the longbow. The traditional shortbow, the weapon of peasants and hunters, was still to be found. But given its ineffectiveness against most armor, it was rarely a critical weapon.

During this epoch, gunpowder was introduced. Initially, it was used for rather crude cannon and soon some not-so-crude cannon. The chief use of these cannon was in reducing fortifications to rubble, which had a fundamental effect upon the conduct of warfare. No longer could an inferior force rely upon its fortifications to hold a superior enemy at bay. It was heavy artillery that was instrumental in bringing about the fall of Constantinople in the middle of the 15th century. Eventually, firearms became small enough to be used by individual soldiers. This did not become widespread until the end of this period. However, it marked the beginning of modern warfare, because no longer was an extremely long period of training required to field an effective fighting man. The age of continuous, high-intensity warfare was dawning.

Mass warfare could still occur on an intermittent scale during this time. In the 13th century, Mongol armies laid waste to most of Eastern Europe and were prevented from overrunning Western Europe only by political events in their homeland. Their army was primarily one of mounted troops, using discipline, archery, organization, and an all-around higher quality "art of war" in general. They easily subdued their less well prepared European opponents. Throughout this period, the Asian cavalry armies were a constant threat to European kingdoms. It was chance, more than anything else, that prevented more encounters. History shows that most of these encounters, had they occurred, would probably have been quite unfavorable for the Europeans.

Meanwhile, the Europeans seemed quite capable of destroying each other. The Hundred Years' War, for example, was a 100-years-plus period of intermittent warfare, pillaging, civil war, and outright mayhem between the English and the French. But it had one positive effect: It helped to make possible strong centralized governments in each of the countries.

There wasn't a great deal of large-scale warfare during this era, primarily owing to the Black Death in the 14th century and the beginning of religious upheaval in the 16th. There was a lot of technical innovation, and you should seek out games that show this.

## THE THIRTY YEARS' WAR AND PRE-NAPOLEONIC PERIOD (1600 TO 1790)

This era begins with one of the most uncivilized wars ever fought in Europe and ends with some of the most "civilized" ever waged.

Between 1618 and 1648, the Thirty Years' War raged. Most of it was fought in Germany. It is alleged, with some substance, that parts of Germany have yet to recover from the hosing they got during that war. It was initially a religious war, but it quickly degenerated into general anarchy, with mercenary armies ravaging back and forth across the landscape.

One of the more important results of the Thirty Years' War was that the damage was so pervasive and intense that there was, by almost universal agreement, an understanding that it was not to be allowed to happen again. And indeed it never has, at least not to the extent of the Thirty Years' War.

As with most wars of such intensity and duration, there was considerable "progress" in military technique and technology. By the end of the Thirty Years' War, there were a number of armies that were more than 50 percent armed with firearm weapons. Pikes, swords, and such were on their way out. By the end of the 17th century, the bayonet was being introduced, and by 1700 the pike was little more than a tradition. After more than 3,000 years of preeminence, the spear disappeared from the battlefield.

Although armies were now armed almost entirely with gunpowder weapons and were somewhat better trained, they were still basically mobs that required hours of stumbling about to get into a battle formation. In other words, battles still had to be fought by mutual consent.

All of this changed with the coming of Frederick the Great in the middle of the 18th century. One of Frederick's most important innovations was his extensive use of the cadence step. This meant troops marching in step in a very highly controlled and disciplined fashion. This enabled him to literally outmaneuver enemy armies and force a battle when the other side might have wished to avoid one. The tactic was relatively unheard of, and for the first time in military history one army could truly control events.

During this period, the standard tactical capabilities of later armies were developed. Indeed, on the tactical level there was very little change during the later Napoleonic era (at least as far as truly effective tactical techniques were concerned). Almost all the weapons that were available during the Napoleonic period were developed during this pre-Napoleonic epoch.

On the strategic level, naval movement ability and ship firepower, as well as the increased ability of governments to marshal financial resources for extended campaigns, was highly developed. And it too went no further during the Napoleonic Wars. The American Revolution, for example (1775 to 1781), was part of what was literally a world war with action going on in Europe and the Far East as well as in North America and the Caribbean. These worldwide operations were largely the result of improved shipbuilding techniques and the arming of these larger ships with many cannon. In fact, there was generally more artillery firepower at sea in this period than was available on land. A fleet of 100 ships (not uncommon) could muster several thousand large-caliber cannon. No land army could match it. Fortunately, the ships were confined to the seas.

It's a pity that there's not more interest among the gaming public for games of the pre-Napoleonic period. This is probably due to a number of reasons, the chief one being that the Napoleonic era overshadows it so much. Yet there was a greater variety of interesting situations during the pre-Napoleonic time than there was during the Napoleonic Wars themselves. There is some evidence of increasing interest among gamers for games of this period. However, this interest must grow quite a bit before publishers can do more games on this era.

Although one publisher has published a game on the Seven Years' War *(Frederick the Great)*, one could make a number of additional games out of this war (fought between 1756 and 1763). The War of the Spanish Succession (1701 to 1714) contained a great number of independent campaigns and some rather striking battles (Blenheim being one of the more famous). The great Northern War (1700 to 1721 in Russia) matched a very interesting group of opponents against one another and the list goes on and on.

## THE NAPOLEONIC WARS (1790 TO 1830)

This period, it goes without saying, was dominated by Napoleon and the French Empire. This was a time of lightning campaigns, national insurrections, the fall of empires, the general reorganization of Europe, and a lot of "color."

What a gamer should look for in this era are many of the tactical and strategic developments that matured toward the end of the pre-Napoleonic period. One of these is the increased importance of leadership. The Napoleonic era was one in which leaders counted for considerably more than in previous times, primarily because there was a lot more good leadership available, particularly on the French side. But the enemies of the French had quite another problem, in that they were saddled with a great number of inferior leaders. This discrepancy had a lot to do with the striking French successes early in the Napoleonic Wars. It wasn't just the ultimately superior numbers of France's enemies that brought Napoleon down. It was more a matter of a slight decline in the quality of French leadership going up against a marked increase in the quality of the leadership among France's enemies.

Many of the so-called tactical innovations of the French were merely temporary expedients. It was the British in particular who had a superior tactical system. When the French met the British, for example, it was quickly discovered that the massed French columns could not bludgeon their way through the disciplined musketry of the British two-man line. So it goes.

## THE AMERICAN CIVIL WAR AND THE 19TH CENTURY (1830 TO 1900)

This was a period dominated by one major war—the American Civil War—and a host of technical, doctrinal, and organizational developments among the many armies that did not fight a major war.

It was by no means a time of absolute peace, only relative peace. The Germans, under the leadership of the Prussians, learned much from their experience during the Napoleonic Wars and developed the effective fighting machine that would be responsible for so much destruction in the 20th century. During the 19th century, the Germans were so superior to their opponents that they humiliated the Austro-Hungarian Empire in a number of days (1866) and took not much longer to smash the French during the Franco-Prussian War of 1870.

The key development was the emergence of the rifled musket. This gave the infantry, for the first time in history, long-range firepower (out to 1,000 meters). This changed (until the 20th century) the whole concept of artillery usage. During the Napoleonic Wars, artillery could be rolled up to within a few hundred meters of the infantry and blast away without much danger of being shot to pieces by the defending musket-armed infantrymen. It was demonstrated with deadly results during the American Civil War what would happen to anybody, infantry or artillerymen, who came within 1,000 meters of enemy infantry armed with rifle muskets.

At the beginning of the American Civil War, most generals on both sides were schooled in the tactics and doctrine of the Napoleonic period. After a number of exceedingly bloody battles, they quickly changed their tactics. But casualties in the American Civil War were considerably higher than in the Napoleonic Wars, even after the new tactical realities had been comprehended. If it hadn't been for the ability of the Civil War armies to maneuver, the American version of World War I would have been fought 50 years earlier. As it was, the lessons of the American Civil War were largely lost on Europeans (most Europeans in power anyway). A gamer who wants to be successful in the American Civil War game on a tactical level cannot afford such blindness.

As noted, outside of the American Civil War there were few campaigns of any great significance during this time. The only exceptions to this general situation were those campaigns involving Europeans against non-Europeans. These were the so-called Colonial Wars. Although the Europeans had generally superior weapons, their tactics and doctrines were not always up to the needs of the situation. Most of these campaigns took place in Africa, although there were some in Asia. Many of them would make quite interesting games, but they too suffer from the same problem that afflicts most campaigns in the pre-Napoleonic era. Most gamers are not aware of them. It's simply going to take time for people to become educated about them. One of these campaigns in particular was noteworthy: the Boer War, fought toward the very end of this period. The Boers, European immigrants themselves, although by now "Africans," gave the British quite a number of setbacks. The earlier campaigns of the Boers and the British against the Zulus also are of considerable interest.

### FIRST WORLD WAR PERIOD (1900 TO 1930)

This period, dominated as it was by the First World War, had a number of previews between 1900 and 1914. There was the Russo-Japanese War in 1905, which on land contained all of the weapons, and results of using those weapons, that appeared during World War I. Not too many people seemed to get the message. After all, they were only Japanese and Russians, and what did they know? In 1912 there was the Balkan War. In fact, there were a number of Balkan wars, which also saw many of the weapons and techniques that dominated World War I put to use. Again, nobody paid any attention.

The basic problem in World War I was that the defense had gained an inordinate advantage over the offense. Rapid-firing artillery, shooting from positions out of sight of enemy infantry (through the use of trigonometry and ground survey), and more modern machine guns, were more than the unarmored flesh of the attacking infantry could withstand. It took the development of armored fighting vehicles (tanks), chemical warfare, and modern infantry-assault (infiltration) tactics to overcome these defensive advantages. Unfortunately, it took four years and many millions of casualties before the solution was found. By then the war was over. Because World War I covered such a wide area, there's no end of interesting campaigns and battles. Immediately after World War I, there were additional interesting wars and battles: including the Russo-Polish War, the Russian Civil War, and various heavily armed disturbances within Germany.

One of the outstanding features of this era, aside from the general destructiveness of the weapons, was the lack of outstanding leadership anywhere. There were a few leaders who rose above the rest, but none of them could be said to have been truly exceptional. It was not a time of "great captains." Any games on the period should show this aspect where it becomes critical, and this is often the case.

At sea, it was the beginning of modern naval warfare. The submarine proved itself in the North Atlantic. The battleships didn't do much of anything, and the beginnings of air power against land forces showed, to those who cared to see, that it was only a matter of time before this new weapon also dominated naval operations.

## WORLD WAR II (1930 TO 1945)

This was the definitive military event of the 20th century. World War I was the introduction, and everything since 1945 has been a variation on the trends established between 1939 and 1945. Although World War II is considered ancient history these days, as the veterans of that conflict become grandparents talking about their distant youth, it still has a powerful attraction for historians and wargamers. Remember that when wargaming first began to emerge in the early 1960s, World War II was *the* war. Korea was seen as a coda to World War II, and Vietnam had not happened yet (and when it did, most citizens, historians, and wargamers wished it hadn't).

Because so many games have been produced on World War II, many of the elements of that war that deserve to be covered by special game mechanics have already been dealt with many times over. In fact, all games on post-1945 conflicts use World War II games as a point of reference. This was not the case with most other periods. By definition, our periods are defined as times when there was no dramatic change. The following period represents the next era of substantial change in how wars are fought. For all practical purposes, we are still in the World War II period. We have defined subsequent periods mainly because World War II was such a vast conflict. This has to be kept in mind when discussing games of the World War II and subsequent periods.

World War II brought about a revolution in the way warfare was conducted, every bit as dramatic as the revolution wrought during World War I. But while the tactical and technical revolution of World War I was essentially static, the revolution and technique during World War II were mobile and dynamic in the extreme. Moreover, the changes emerging in World War I were only the preliminaries for the full-blown developments of modern-warfare techniques that matured during the 1930s and World War II.

For the first time in history, there were armies consisting of millions of fighting men going at one another over a battle line literally thousands of miles long. It was possible to kill hundreds of thousands of soldiers in the course of a weeklong campaign, while the victorious side might advance hundreds of miles. This last element, the huge territorial gains, was what distinguished World War II from the more static World War I. The basic ideas of mass warfare were around during 1914–18, but the mechanization that put the "blitz" in the blitzkrieg was not available until the 1930s.

To fully appreciate the implications of all this, one must realize that this type of dynamic activity was in the past restricted to battlefields usually no more than three or four kilometers in diameter. Until trucks and tanks were developed (and mass-produced), troops could only advance as far as they could walk. Horses never really provided the same kind of mobility, as horses required enormous amounts of food and were very prone to injury and disease. Firearms made horses more vulnerable to injury on the battlefield than humans. With mechanization, the same fluidity and dynamism of movement found in smaller battlefields was now being played out on battle zones hundreds of kilometers in diameter. If nothing else, this made maps of the battle more recognizable to the average person since they were maps of entire regions rather than small-scale maps of obscure crossroads where armies happened to meet one another.

Game designers didn't have too much trouble handling land operations in World War II. It wasn't much of a jump going from the Hussar squadron of 100 or so men on horseback to the Panzer division of 12,000 men (and more than 2,000 motor vehicles). Artillery had increased in its range, but then so did the scale of the game map. In a Napoleonic battle, each hexagon on the game map represents 100 or 500 meters. In a World War II battle where the artillery can fire 10 or 20 times farther, the hexagons represent 10 or 20 times as much space.

In land warfare, the unique dynamics of mechanized combat have been treated in an extensive variety of game rules. The primary thing the gamer is trying to re-create here is the ability of mechanized forces to rapidly concentrate on a portion of the enemy line, make a breach, then exploit it by pouring a certain number of highly mobile mechanized forces through it. This mechanized group then surrounds large portions of the enemy forces. What I have just described is what is basically known as the "blitzkrieg." It was still being used in its World War II form as recently as the 1991 Desert Storm operation.

Where things really get complicated in World War II games is where you start simulating the weapons that never existed (in a big way) before 1939. Take air power. This spawned any number of problems for the game designers. You now had things such as airborne infantry, troops that landed by parachute. They had airplanes flying off ships. Those aircraft carriers had to be dealt with. Although ships still looked like the ships of fifty years before, dramatic changes had taken place. Entire armies consisting of hundreds of thousands of men could now be landed from the sea within the space of a few days. Thousands of aircraft could be put into the air against one enemy target within a few hours. Over the past ten years, most of the design problems have been solved in one fashion or another. There is still considerable room for better solutions, and much of this work will be done not only in the World War II games but also in the games of the contemporary era.

Gamers should be looking for one of two things in a World War II game. Both of them are opposite extremes. For those gamers who want a fairly simple game, they should look for one that deals "elegantly" with the key problems in the particular situation being simulated. At the other extreme are the more detailed games, in which the player is challenged to find more innovative and effective solutions to the unique problems occurring in this particular period.

Since so many of the World War II battles have been done quite often on the same scale by different publishers, one is beginning to see some very elegant treatments of the more popular battles. This is creating extremely interesting games. Although the World War II era has lost much of the popular appeal it once had, it is still probably going to be the era from which some of the more interesting games will come. And in the final analysis, World War II is still the area from which a disproportionate number of new ideas and concepts will come. The era is just too big. Too much was going on, and it's too recent to be ignored as many of the older periods are. Hobbyists will dash off into the future and into the past for the novelty of it all, but they will continue to come back to World War II for the meat and potatoes of simulation in gaming.

### GAMES OF THE CONTEMPORARY ERA (1945 TO A FEW YEARS FROM NOW)

Games in this area fall into two categories: those that are already historical and those that haven't happened yet. Most of the historical games in this period (from World War II to the present) are no different really from games done on World War II or an earlier topic. They are, however, more topical, not so much because they are closer to our present time, but because the developments in warfare technology that they demonstrate are that much closer to what would happen if certain armies went to war today or tomorrow. This is an important consideration when dealing with games of the present and immediate future. What all these immediate future and present games are can be summed up in one word: extrapolation.

Extrapolation is nothing more than taking a current trend and continuing it into the future. This works quite well in historical research. No technique is perfect, but extrapolation has served better than any other. Of course, the farther into the future one gets, the more shaky one's extrapolations become. This makes life a little more interesting. It also makes observing many of the contemporary period games interesting. When the events they "predict" actually occur, it's interesting to see how close they were to what could and did happen.

Understandably, people in the military are most interested in these "predictive"-type games. Not only the military but also many civilian and nonmilitary governmental organizations have a keen interest in games of this sort. "Why," you might ask yourself, "would all of these organizations with all of their resources be interested in a predictive game from a civilian organization?" The reason is very simple. Civilian publishers do their games out in the open. They are published for all to see and criticize. As one book analyzing the professional wargaming scene put it (*The War Game,* Harvard University Press, 1977, long out of print): "Some of the planning factors used in amateur wargaming may even be more accurate than those used by the professionals; at least the data are more openly available and are actively challenged by this large and active group of amateurs."

Games on wars not yet fought have proven remarkably accurate. I designed *Sinai* during the summer of 1973, only to see my predictions for the "next" Arab-Israeli war come true. In 1990 I worked with designer

Austin Bay on a Desert Storm game. It was published in December 1990, and accurately predicted all that happened in the next two months. We are dealing with some powerful predictive techniques here, and the techniques are powerful because they are constantly honed on historical situations. Beware, however, of predictive games. The difference between a false prophet and a real one is usually detectable only after it's too late. Proceed with caution.

### FANTASY AND SCIENCE-FICTION GAMES

These games, of course, are based upon fictional subjects. The science-fiction games are usually founded upon books, stories, or, at the very least, ideas, as found in contemporary science fiction.

Fantasy games pay much less attention to science and much more attention to magic. I should point out, however, as a historian, that the plots and general frameworks of almost all of these games are derived from ancient and medieval topics. Much science fiction postulates an era of exploration, discovery, and facing the unknown. It sounds like Christopher Columbus all over again. Fantasy games will also follow those plot lines, but in addition will delve frequently into ancient and medieval mythology and religion. Gamers I've surveyed who had a strong interest in fantasy and science fiction had an equally strong interest in the medieval period. This was surprising at first, but upon a little reflection not so surprising at all. This may explain how users of even fantasy games can get into arguments about "realism." Fantasy is often little more than the past the way we would like it to be. Science fiction is the future the way we want it to be.

Computer fantasy and science-fiction games are very popular because they can make visual things that otherwise are unlikely to be seen. With historical games, we have a wealth of pictures from the past. Fantasy and science fiction are largely invention, and computers allow the invented visuals to be made concrete.

Unlike historical games, fantasy and science fiction have fewer restraints on what they can get away with. The designers as well as the users are eager to try anything. I find many innovative ideas concerning game mechanics can be found first presented in fantasy and science-fiction games. These ideas can then be applied to historical subjects.

May the force be with you. So it goes.

## SPECIAL PROBLEMS OF AIR AND NAVAL GAMES ✱✱

Without belaboring the obvious too much, I should point out that most gamers, when forced to choose between games taking place on the land, in the air, or at sea, will tend to prefer the land games. Naval comes next, a distant second, and an even more distant third are air games.

One would think that naval games would have a certain popularity, given the long history of naval warfare. Much of the history of naval warfare that people are aware of is a myth. Most naval warfare in the past (and to a certain extent in the present and future) is rather dull and prosaic. Up until about 200 years ago, most battles consisted of large groups of ships sailing or rowing into one another and then proceeding to attempt to ram or burn down (or even better yet capture) the opponent's ships. This was not a very elegant form of combat. And it was even less controllable than similar forms of combat found in land battles.

These shortcomings can be got around somewhat by having a game that concentrates on the operation of single ships. The individual ships certainly had an interesting situation before them. They could maneuver, and although they might have only one opportunity to ram or lash themselves beside an enemy ship, the talents of a skillful seaman could be most decisive.

During the latter half of the 18th century, naval fighting tactics were developed that put a premium on organization and coordinated movement and action. The British were the first to seize upon this idea in an organized fashion, and this was a chief factor in "Britannia ruling the waves" for the next 150 years. Even with this new organized way of combat, the ships involved were still sailing ships and were heavily dependent upon the vagaries of the climate and the weather. Individual skill and seamanship were still paramount.

During the middle of the 19th century, this changed with the introduction of steam propulsion. Shortly thereafter, the armor plating of ships really got going because the vessels now had the power to lug around all that weight. This was not the case when metal plating was first introduced during earlier periods.

The introduction of the wireless telegraph early in the 20th century again revolutionized naval warfare since it allowed coordinated strategic operations. Prior to this, fleets would be sent off thousands of miles on missions of, it later developed, dubious worth. Yet these missions had to be completed since there was no way of recalling the fleet or informing it as to what was going on. Particularly during the 17th and 18th centuries, sailing squadrons and fleets were sent all over the world. This led to some curious battles being fought after the war was "officially" over.

The introduction of the submarine and the aircraft carrier drastically changed all previous concepts of naval warfare. Up until World War I, naval power was still coming out of the barrel of a gun. With the advent of the submarine, considerable power shifted to torpedo tubes. As the war and merchant fleets had increased in size considerably with the introduction of steam propulsion, so nations became much more dependent upon the trade that these merchant vessels carried and warships defended, or threatened. Submarines quickly loomed as the most serious threat to that trade. Naval warfare now became a battle between unequal antagonists. Submarines were not really equipped to take on surface warships, yet there were never enough surface vessels available to track down all of the submarines.

With the coming of aircraft, the submarine was somewhat neutralized. Nuclear-powered submarines changed a lot of that from the 1960s onward. But all ships were now even more vulnerable to destruction by aircraft. The aircraft carrier became one of the chief means of this destruction.

The importance of electronics and "electronic warfare" also increased evermore. Air warfare, more than any other form, quickly became dependent on electronics, radar, and antiradar devices. Air navigation depended on electronic emissions. Even the weapons the aircraft used quickly became creatures of electronically controlled instruments.

Aerial warfare itself is rather mundane. The vast majority of combats in the air follow the "Theory *X*" concept of battle. That is, one side ambushes the other. When two aircraft spot each other simultaneously without one having the jump on the other, the combat is usually inconclusive. Current theory, however, holds that the lethality of electronically guided air-to-air weapons is such that two aircraft, upon spotting each other, will almost automatically destroy each other. Given the history of self-preservation among combatants and the recent history of those aerial combats that have occurred, this theory of mutual destruction is likely to be questionable.

As with naval warfare, aerial warfare can be most interesting and illuminating when conducted on the level of individual aircraft. Many games do just this. This is particularly the case with the most popular form of computer wargame: the aircraft simulator. These simulators never succeeded in manual form; there was simply too much detail for the player to keep track of.

Naval games have also become relatively more popular in their computer incarnations. Again, the highly technical aspects of naval warfare can more easily be handled with a computer.

# 4 DESIGNING WARGAMES

Game design is very much like writing a book, term paper, or any other work of nonfiction. In many respects, it's actually easier. But in some respects it's definitely different. In this section, I will explain the differences.

The main difference is the structure of the work. A nonfiction writing assignment is basically an act of communication. The writer collects, reorganizes, and presents data in a form that the reader can easily use. This type of communication is what I call "linear." That is, we start at the beginning and proceed word by word, sentence by sentence, paragraph by paragraph, page by page, along the line laid down by the author. A game, on the other hand, is nonlinear. To be sure, it starts at a beginning, but from there on it is primarily an exercise of choices and options, of different paths that the reader or gamer may follow. Designing a game is working with a much more structured piece of work than is writing nonfiction. For one thing, a game is much more graphic than a literary work. A game is much more precise. Even on a "literary" level, it is mainly a set of instructions. But minor mistakes or ambiguous passages that might not seriously harm a nonfiction work can cause grave problems in a game.

There are many rules to designing games. Above all, there are two game-design rules that control all others. First, and most important is:

*Keep It Simple.*

The second rule is nearly as important but is a bit more complex in its use. The second rule is:

*Plagiarize.*

Plagiarism is a dramatic way of saying, "Use available techniques." If you try to plow too much new ground, you're not going to get very far, and you will have an extremely difficult time in keeping your game sufficiently simple to be manageable.

It is very difficult to keep a game-design project simple. Once you get going, there are tremendous temptations to add this and that. A game design is a very dynamic activity. It soon acquires a life of its own, asking questions and providing parts of answers. The game designer is sorely tempted to go deeper and deeper. Without some years of experience and a high degree of professional discipline, it is extremely difficult to do an unsimple game that is not a truly incomprehensible one. For a game is, in addition to being a source of information, also a form of communication. If the information cannot be communicated, the game does not work. You've got to keep it simple.

Using available techniques provides you with a wide range of proven procedures on how to design a game that gives you the most bang for the buck. I assume you only have limited time for designing a game, and I presume you would prefer to spend less of your time banging your head against a wall and more time refining a functioning game.

There are no iron-clad, guaranteed ("Follow these rules and you cannot fail") guidelines for designing a game. I do have, however, 10 steps that are generally followed in successful wargames design. The number of steps has actually changed over the many years that I have been designing games. But, in essence, these 10 steps, albeit rearranged and renamed occasionally, have remained remarkably consistent.

**1.** The first, and most important step, is concept development. You must determine at the very beginning what it is that you want to do.

**2.** Next comes research. You will have done a little of this during the concept-development stage. At this point, you must fill in as many of the gaps in your knowledge as you can.

**3.** This is what I have dubbed "integration." This is where you take all of the research material and your knowledge of game mechanics and integrate it into a prototype game.

**4.** Now, you flesh out this prototype, coming up, in effect, with something that looks remarkably close to the finished game. In some cases, if you're lucky or the phases of the moon happen to be just right, your prototype will be exactly like your finished game.

**5.** Prepare a first draft of your rules. Many people overlook this step, preferring to keep the rules in their heads for a while longer. They usually come to regret this.

**6.** This is one of the more difficult steps: game development. This means play-testing and changing the game and rewriting the rules and taking a lot of abuse from people who would rather play than design and don't appreciate at all the problems the poor designer has in getting anything done.

**7.** I call this step "blind-testing" (computer-game designers call it beta-testing). This is where you take your physical prototype and your written rules and send them out to somebody who can play the game without your presence. This is often very revealing.

**8.** Editing. This step occurs when all of your blind-testing results have come back and have been integrated into the manuscript. Somebody else should now take over the manuscript and edit it. This also means trying to play the game with all of your final corrections and changes and generally attempting to smoke out as many gremlins as possible.

**9.** Production. If you are going to publish the game, this is the production step. This is where many things can go wrong. Rules have to be typeset, and things can get scrambled about. The artwork has to be prepared from your prototype, and things can get changed again. There is much potential danger in this phase of game design.

**10.** Feedback. This step is also extremely critical if you are going to design any more games. This is the step where you must systematically collect feedback from those who play your game to see where you went right and where you went wrong.

These steps are used for the publication of historical games. Most people will apply their game-design skills to the modification of existing games or to the design of games that will most likely never be published. For these unpublished games (I deliberately refrain from calling them "amateur" because I am consistently impressed by the quality of unpublished games compared to many of those that are published), the designer's effort should be directed toward producing a game that he can easily and effectively use. This means that while all of the 10 steps are used, blind-testing, editing, and production tend to be folded back into development and prototype construction. But it would not be that difficult for a gamer to conduct a blind test as well as having some other player edit it and then going through some simple production procedures that would enable the gamer to give out copies of his game without going to the expense of actually publishing it. I will explain these techniques further on.

There is a lot more to the 10 game-designing steps than I have briefly explained here. The tricks of the trade are what make these steps functional, tricks that have been uncovered in the last 30 years and could fill a few books. In order to get across to you a good number of these tricks in a format that you can use, I will describe the actual design and development of a game that was prepared expressly for this book. Although a rather small game, it has all of the same elements that much larger games contain and requires that the designer go through all of the same steps and encounter and solve the same problems.

## DESIGNING A GAME STEP BY STEP ✱✱

The subject chosen for the game in this book was the attempt, in 1944, of General Patton's Third Army to penetrate the German lines around the old fortified French city of Metz and thus reach the Moselle River. Doing this in mid-September 1944 would have had the effect of collapsing the entire German defense effort west of the Rhine River. The losses the Germans would have then suffered in retreating behind the Rhine with their non-mechanized forces would have made it possible to end the war five or six months earlier.

Now, that's the description of the situation being simulated. It was not chosen just for its historical interest, although that was one important reason. A more important reason was that the situation could fit on a five-by-eight-inch game map. In addition to the small size of the situation, it is a relatively simple one. The scale of the game is such that regimental-size units are used, thus giving eight American units and eleven German. The basic concept of this game, then, is a teaching game, one that will fit in a book and that will have sufficient simplicity to easily demonstrate various aspects of game design.

It is important to remember that in any historical game you basically have two concepts you must adhere to. One is fixed, and that is the concept that the historical event must be accurately simulated. The other is more variable and has to do with in how much detail you wish to portray the historical event. The different levels of complexity available in a historical situation strongly affect the game's accessibility. A more complex game may give the user much more, but at much greater expense in terms of his time and effort. It's the difference between having a full dinner and a snack. Many gamers prefer to approach a large number of game subjects as they would a smorgasbord—if only to determine which historical period or event might excite their interest to the point where they will be willing to expend the great amount of time and effort required to deal with a much more complex simulation on that particular subject or area.

This accounts for much of the popularity of simpler games. Indeed, quite often the same gamer who will spend dozens of hours playing a complex and detailed simulation on a particular situation will also be found playing a much simpler simulation on the same subject when time and energy are short.

The concept stage of game design also allows you to decide what particular aspect of a situation you wish to spotlight. You may have a pet theory regarding the critical elements of particular historical conflict. You can design your game to emphasize these particular elements, be they leadership, weather, geography, the effects of certain weapons, or whatever. Generally, the more accurate your perceptions regarding the critical elements in the battle, the better the game will be. But even if your stress in the game is at variance with the real critical elements in the situation, you still end up with a rather interesting game. It's simply a question of what aspect you will be viewing the historical situation from.

Now that we've decided what we want our game to be, we have to put it together. This is the second step of designing a game, the research step. It was your knowledge of the situation that led you to think about doing a game on it in the first place.

The sources that will go into the *Drive on Metz* game number about a dozen volumes in all. These include the volumes that must be resorted to for technical details of things that go into the game. But the main source, the one that is followed to determine the faithfulness of the game to history, is *The Lorraine Campaign* by Hugh M. Cole. This is an exceptional book in that it contains information on just about every aspect of the original Metz conflict. This is unusual. It's no accident that *The Lorraine Campaign* is put together this way, as it is one of the volumes in the U.S. Army's officially sponsored series on the history of World War II. Indeed, the book was published for the U.S. Army by the Government Printing Office.

The book contains rather complete Order of Battle information, excellent maps, and the results of additional research that army historians did through interviews with German participants in the battle. I've worked in the army archives in Washington; not only is the research there done from the "enemy point of view" invaluable in doing a game, but the American Army did an admirable job in incorporating much of this information in its official histories. Indeed, the only fault one can find with the army's official histories (especially when compared to those done by European countries) is that they do not include real nuts-and-bolts information on the organization and real abilities of the equipment and units. This is covered to a certain extent by other volumes in the U.S. Army series, particularly those involved with the procurement and training of ground combat troops. But for a game of the scale of *The Drive on Metz, The Lorraine Campaign* is quite adequate.

Although *The Drive on Metz* is a simple game, this simplicity was achieved only through the use of extensive knowledge of the event and period in which it took place. The following bibliography consists of additional sources that contain extensive information:

*The Lorraine Campaign* (Hugh M. Cole, Government Printing Office). This volume is part of the U.S. Army's official history of World War II. It is a thick book, unbiased, and goes into considerable detail on the campaign that the Drive on Metz belonged to. Basically, though, this source was a detailed diary of the battle.

*Handbook of German Military Forces* (U.S. Army Intelligence). This was an official document prepared during the war and was the most complete single source for details on the German Army (organization, tactics, weapons, etc). There are usually reprints of this book available in specialized bookstores.

"The Organization of the U.S. Army in Europe (1944–45)" (*Strategy & Tactics* magazine #30). This article contains organizational, tactical, weapons, and other information on the U.S. Army.

*Arnaville, Altuzzo and Schmidt* (MacDonald and Mathews, Government Printing Office). Another of the U.S. Army official histories. This is one of the specialized volumes, in this case a detailed study of three battles in Europe. One of these battles (Arnaville) took place during the Drive on Metz. The section on Arnaville gives much detailed information on the entire period covered by the *Drive on Metz* game.

I could go on with bibliography related to the game. As one designs more games, one draws upon more information. The above sources, however, would give the average person 95 percent of the data needed to design a Drive on Metz game. The other 5 percent is hard to put on paper.

Our next step is to make our game map. This is done very simply. On page 15 there is a black-and-white reproduction of the full-color map taken from the many maps in the Lorraine Campaign book. The area covered by our game is outlined. This map is then overlaid with a hexagonal grid (we usually do this on a light table, putting the map under a blank hexagon sheet and using the light shining from underneath to simply trace the terrain features; you could use a window on a bright day for the same effect). The next map is the play-test map (again a black-and-white rendering of a color map). And finally we have the finished game map. In this case, the map was done to be reproduced in black-and-white. Although color helps, you can see from this map it is not absolutely essential, and the terrain features are clear, distinct, and capable of being played upon. The terrain analysis was not difficult at all. Indeed, the following excerpt from *The Lorraine Campaign* coupled with a reference (hexagon numbers) as to the map in question will demonstrate how simple it is:

**THE XX CORPS'S CROSSING OF THE MOSELLE**

The military value of the Metz position lay not in the size of its garrison or in the intrinsic strength of its numerous fortified works. Instead, the long defense of Metz must be ascribed to a combination of factors favorable to the Germans: the presence of elite troops during the initial stages of the battle; the moral and physical strength derived from steel and concrete, even in outdated fortifications; and the possession of ground that favored the defender.

The eastern face of the Meuse Plateau, whose heights average some 380 meters, falls sharply to the plain of the Woevre and a mean elevation of not more than 220 meters. In this plain the Imperial German armies had deployed for the bloody frontal attacks against the Verdun salient in 1916. Beyond the Woevre the Moselle Plateau rises gradually to command the western approaches to Metz. The western edge of the plateau coincides roughly with the Conflans-Mars-la-Tour-Chambley Road [hexes 0206-0207-0208-0209]. The eastern heights, averaging 370 meters, drop abruptly to the Moselle River. East of the river some blocks of the Moselle Plateau reappear, but these are dominated by the higher ground on the west bank. The main plateau, if measured from Conflans to Metz, is about 10 miles in depth. The western half is moderately rolling; on some roads the ascent to the east is barely perceptible. The eastern half of the plateau is high, rugged and wooded, grooved by deeply incised ravines and innumerable shallow draws. It would be hard to imagine a terrain more compartmentalized and conducive to defense by small tactical bodies.

The Metz salient, as it confronted the XX Corps at the beginning of the September operation, extended for some 18 miles in a perimeter west of Metz and the Moselle. On the left the German position rested on the Moselle near Arnaville, about nine and a half miles from the center of Metz. On the right a western affluent of the Moselle, the Orne, marked the limits of the German line, which was anchored near the village of Mondelange [hex 0103] approximately 10 miles due north of Metz.

At the southern end of this bridgehead position, the ravines cut obliquely through the wooded Moselle scarps and the defile down to the river channel. The Rupt de Mad, farthest from Metz, is traversed by a road that angles from Mars-la-Tour via Chambley and reaches the Moselle near Arnaville [hex 0510]. The middle road riverward can be entered either at Mars-la-Tour or at Rezonville. It then passes through the village of Gorze [hex 0409], lying in the main throat of the gorge to which it gives its name, and attains the Moselle bank at Noveant. The third and northernmost of these ravines, the Mance, forms an "L" whose upright runs from north to south through a small depression in the Bois des Genivaux. Near Gravelotte this shallow gully descends into a deep draw, finally turning toward Vaux and the Bois des Ognons. Just east of Gravelotte the main high road between Verdun and Metz dips to cross the Mance, while a secondary road [hex 0508] branches south at Gravelotte and follows along the bottom of the ravine to Ars-sur-Moselle and the river.

These three defiles would canalize any attempt to turn the Metz position on the south by a drive to and across the Moselle. But a close-in envelopment or a frontal attack in this section would be hampered chiefly by the ravine of the Mance. In effect, therefore, the natural anchor position on the German left was formed by the lower Mance ravine, the plateau of the Bois de Vaux north of the ravine, and the plateau of the Bois des Ognons to the south. On the eve of World War I the German governors of Metz had reinforced this natural abutment by the construction of a heavily gunned fort on the river side of the Bois de Vaux Plateau about a mile southwest of Ars-sur-Moselle. This strong work, renamed by the French in 1919 as Fort Driant, was sited so that its batteries dominated not only the southwestern approaches to Metz but the Moselle Valley as well.

North and west of the Bois de Vaux two villages, Rezonville [0407] and Mars-la-Tour, served as outpost positions for the southern sector of the German front. They blocked the main road to Metz and controlled passage from north to south through the Mance and Gorse ravines. Beyond Gravelotte, the Bois des Genivaux and the wood-bordered Mance combined in a strong defensive line and masked the German forts farther to the east. These rearward positions lay on the open crest of a long ridge whose western slopes were outposted by a sprinkling of isolated but strongly built farms.

North of the Bois des Genivaux, the forward German troops occupied a plateau marked by the villages of Verneville [0406] and Abbeville. The strongest position in the German center, however, was farther to the east. Here the village of Amanviller, located on a tableland, lay under the guns of forts hidden on wooded ridges to its rear. The Amanviller Plateau continued northward on the German right. In this area the forward defense line included the villages of St. Privat and Roncourt. To the rear rose a welter of rugged heights and heavy forests, running diagonally northeastward to the Orne. This northernmost portion was held only lightly. The main German line was a kind of switch position extending from the Bois de Jaumont along the Bois de Feves ridge. This switch position was strengthened by a series of forts and walls. In this sector, however, the Moselle scarps do not come clear to the Moselle, as they do south of Metz. Here, in the area of Mazieres-les-Metz, a wide, level floodplain offered a gateway to the Metz position, once an attacker had cleared the western escarpment.

In sum, the ground west of Metz gave a very considerable advantage to the defender. Long, open slopes provided a natural glacis in front of the main German positions. Wooded crests and ravines screened the movement of troops and supply from the eye of the attacker. Broken terrain permitted the

> most effective use of small defending groups. Ravines, draws, and thick wood-lots offered ample opportunity for counterattack tactics, both in force and in patrol strength. Finally, the German soldier had used this terrain as a maneuver arena and was prepared to exploit every accident of ground.
>
> —From *The Lorraine Campaign* by Hugh M. Cole

Now we get into the third step, which I call integration. Here we integrate this research material into game components. I am also jumping right ahead into the fourth step, which is preparation of the prototype. Indeed, at this point we can even stretch out into the fifth step, which is preparation of the rules. I will explain how I can be stretching ahead of myself like this by describing the tools available to a professional game designer and how they will, if properly used, produce a highly effective game with a minimal expenditure of resources (not to mention decreasing the migraine-headache count for both players and publishers). Once you've got your map, the next thing you do is integration in the truest sense of the word. You have to put together your Order of Battle, or your units that will be represented as playing pieces in the game, as well as your Terrain Effects Chart and, to a lesser extent, your Combat Results Table.

The playing pieces contain information relating to their combat abilities (combat strength) as well as their movement ability (movement allowance). Both of these numbers are subject to change as the game is developed, but you must put down on the Terrain Effects Chart and the Combat Results Table some numbers to start out with. For *The Drive on Metz,* I did it like this. The scale of the game map is 1:250,000 which comes out to four kilometers per hexagon. A unit moving through a hexagon actually moves farther, since one does not move in a straight line. So the actual distance in moving through a hexagon would be closer to five or six kilometers. Most of the units are infantry divisions. This infantry is not motorized, and, without special transport being attached to the division, the infantry marches on foot.

Most of the rest of the division is motorized, particularly the heavy equipment and the units that carry supplies, but the division as a whole can move no faster than the infantry can march. Moving on good roads, you can expect the infantry to move about four hexes or 16 kilometers a day. Motorized units, despite the fact that nobody is walking, are not that much more efficient if only because of the sheer size. Motorized units are also more limited in a tactical situation by the fact that a division does contain a couple of thousand motor vehicles and the road net is never dense enough for all the units to move everywhere they want at the same time. All of these problems conspire to give motorized units no more than two or three times the movement ability (in most situations) of nonmotorized formations. This being the fact, we assign the movement allowances to the various formations. The Germans get a somewhat lower movement allowance for their motorized units because their equipment is not up to the standard of the Americans, and the American air power has by this time developed a rather nasty habit of blowing away German motor transport with sickening regularity.

Getting back to the Terrain Effects Chart, we thus have the road movement requiring one movement point, the clear areas requiring two, since the basic assumption is that it's more difficult to move through fields (no matter how clear they are) than to move on a road. Rough terrain simply represents more slopes and rough terrain in general. The forest actually is the most difficult terrain to move in, especially for the motor vehicles. It can be done, but very slowly.

Calculating the effects of terrain on combat is a bit more difficult, but as with most problems encountered in game design, it is best to approach it from a rather simplistic view. First of all, the Combat Results Table is constructed on the assumption that all combat will take place in fairly open terrain. Thus, all of the effects of terrain on combat have to do with improving the position of the defender. On our Combat Results Table, different types of terrain reduce the attacker's odds by the simple expedient of moving one or more columns to the left on the Combat Results Table (a +5 becomes a +4 or a +3 depending upon the terrain the defender is in).

Terrain has two advantages for the defender. It will hide him, thus making it easier for the defender to ambush the attacker. Don't underestimate the value of ambush. Most casualties in warfare, especially modern warfare, are nothing more than successful ambushes. For the defender to successfully ambush the attacker, he must be capable of concealing his presence. Granted that the attacker may know that the defender is in the general area, but any battle comes down to those individuals up front making contact with individual enemy troops up front. A forest, a town, or exceptionally rough terrain is not a welcome sight to a soldier, mainly because of what he cannot see. The second advantage conferred by terrain is protection from enemy fire. Some types of terrain will conceal you from sight but not from the effects of modern weapons. Thus, a forest can be a mixed blessing. Indeed, many types of forest are more dangerous when hit by artillery fire because the exploding shells create additional lethal wooden fragments when they blow up trees. You're just as dead from a tree crushing your skull as a bullet wound. A town should, ideally, be the best type of terrain to defend in. However, as can many other types of terrain, it can aid the attacker also. Once the attacker gets into the town, he's covered (from the effects of enemy fire), and concealment works both ways.

Rivers, streams, and similar bodies of water are a special case. Their main effect upon the attacker is to ''embarrass'' him. That is, the attacker must first of all appear on the bank of the river and then (never quickly enough) attempt to build a bridge or use boats to get across. At the same time, the defender has a golden opportunity to kill all of the attackers before they can get to the other side. This often happens. The only way the attacker can overcome it is to use his artillery and long-range-fire weapons to reduce the effects of the defender's fire as much as possible. That being the case, a frontal attack across a river is always a dangerous operation.

The fortified hexes on the game map are the most important for the Germans. They represent, as the name implies, fortifications. These complexes consist primarily of artillery and machine-gun positions built into concrete buildings covering areas from a few hundred square meters to a couple of acres, usually occupying the high ground and having open areas around them to make it difficult for an attacker to approach them without being seen. Since these forts generally block the movement routes through many key hexagons on the map, they pose a major obstacle to an attacking force. However, these forts are of use to the Germans only after they are occupied (a German unit in the fortified hex).

So there's our map and, coincidentally, the Terrain Effects Chart. The next step is to put the playing pieces on the map.

First, we have to construct the Order of Battle. This is initially nothing more than a list of the major formations that took part in the battle.

The scale of the game generally determines what size the units represented in the game will be. The battlefield for our game covers an area 44 kilometers wide by 36 kilometers deep. During World War II, it was normal for a division (the basic unit for ground combat for the last 100 or so years) to cover a frontage of 10 to 20 kilometers. However, it was not unknown for divisions to stretch themselves out over 50 or 60 kilometers. Since a division-size game unit, with its adjacent Zone of Control hex, could only cover three hexes (12 kilometers), we have to take a look at the next level.

In the Drive on Metz, there were four divisions on the German side and three on the American side. Note that this is too few units for a ''normal'' game. Generally, you need about 10 per side to make it interesting. That is, provide sufficient decision-making opportunities for both sides.

Since each division consists (usually) of three regiments or brigades, we achieve, by this simple piece of logic, the decision to use regiments in the game. Regiments and brigades normally cover about half a division's frontage. One third of a division or any other unit is kept back as a reserve, if possible. Thus a division holding a 20-kilometer front would put two regiments up front, each holding 10 kilometers, and keep the third regiment back as a reserve. This solves our problem. If we use regiments, they will be covering as much terrain (12 kilometers max) and provide a sufficient number of units (about 10 each side) to make the game interesting. Each playing piece will represent one of the regiments that participated. The next thing we must do is to make up a list of all the regiments that did, indeed, take part. This is usually pretty straightforward (although for some battles it can be a real headache), and on this list we should simply note whatever information we can obtain on the quality and quantity of each of the units. In our Metz game, for example, subunits of the three American divisions (the 7th Armored, 5th Infantry, and 90th Infantry) were all fairly equal, using the standard organization then prevalent in the American Army. The 5th Infantry was a bit more experienced, and the 7th Armored, because of its equipment, was a bit more powerful as well as being completely mobile. Each of the American regiments consisted of (with all attached and supporting units) some 3,000 to 4,000 men. On the German side, it was a somewhat less uniform situation.

The Germans had four divisions available, but one of them wasn't really a division. Two of the others were only remnants of divisions. The "complete" division was the 559th Volksgrenadier, which consisted of recently (albeit, well) trained but green troops in one of the new divisions the Germans had organized since the Normandy invasion three months before. In terms of manpower, the regiments had a bit less than two thirds of that available to American regiments. In addition, the German units (if not many of the senior individuals) lacked experience relative to the American units and, more important, did not have the lavish levels of equipment and ammunition available to the Americans (even with the American supply difficulties of the time).

The German "nondivision" was Division Number 462. It was referred to as that because it was simply a divisional staff (a few hundred officers and clerical staff) whose main function was to gather under their control whatever units that could be remotely considered combat-worthy and to attempt to function as a division. Actually, Division 462 did rather well in this respect because it had available to it in the area of Metz the staff and students of two training establishments, one for officer candidates (the regiment Fahnenjunker) and the noncommissioned officers' school (the regiment Unterführer). The Fahnenjunker Regiment had a strength of 2,200, but mediocre equipment. The Unterführer Regiment had only 1,600 and about the same level of equipment. The 3rd regiment of Division 462 was the 1010th Security Regiment, a unit that just happened to be in the area. This unit was reduced to a level of 600 second-rate troops, but they were fairly competently led and could hold a position for a while. Finally, there was the Metz garrison, consisting primarily of support troops in the area of Metz itself. These troops were not very mobile, but there were a few thousand of them, and they could provide some resistance if it came down to actually defending Metz.

The two remnants of divisions were actually of rather high quality. There was the 17th SS Panzergrenadier Division and the 3rd Panzergrenadier Division. Each of these divisions had only two regiments (which was the way they were organized) and each of these regiments could muster only a few thousand men at most. In effect, these divisions were now two segments, each of its two regiments. The divisions had been battered about considerably in retreating across France during August, but their morale was still good, and they were starting to rebuild. Indeed, the 17th SS had just incorporated into its ranks two SS brigades (the 49th and 51st) that were being sent forward to reinforce the crumbling German front. The 3rd Panzergrenadier had come up from Italy, where, although it had been punished a bit, it had remained in pretty good shape. Finally, the Germans had one reserve unit available to them during the battle. This was the 106th Panzer Brigade—a few thousand men with good tanks and other equipment, but very poorly trained and led. This unit did not distinguish itself.

Assigning combat values to these units is a fairly simple process initially. Simply take the worst unit (as best you can determine it) and assign it a value of "1." It's a good idea at this point to take the best unit (as best you can determine it) and ask yourself the question: How much better is the best unit than the worst unit? If you come up with a number no greater than, say, 9, you're probably in the ballpark, and at that point all you have to do is fit all the remaining units in the game in between the best and the worst by just asking yourself the question: How is this unit in comparison with the best and the worst? If it sounds simple, it is. There's no mystery involved in it. The system is further refined when you start playing the game, and any misjudgments you have made quickly become evident in your attempts to re-create the historical event. You then modify the values on units and eventually end up with a rather accurate numerical appraisal of each unit's combat ability.

We now place the units on the game map. Simply put, the units will start the game in one of two ways: Either they will be placed on a specific hex at the beginning of the game, or they will enter the game map from a specific side (or even hex) where they originally entered the area. In this particular battle, the German units are placed on the map where they were historically on September 7, and the American units enter as they did in early September.

At this point, it would be a good idea to explain the concept of "level of abstraction."

On September 7, when our game begins, there were actually American units on the map, but not units in the sense of the game. These were very small formations, primarily reconnaissance patrols. Since we are not dealing with a game that includes units as small as half a dozen men and two jeeps (a reconnaissance patrol), these units will not show up. They will be "abstracted" when the "main force units" enter the map. The small recon unit will be assumed to be wandering in front of the playing pieces that are used in the game. This is a common and necessary practice in designing games. It is not possible to include every element, no matter how complex the game you are trying to design. When you are designing a game, you must keep firmly in mind what level of abstraction you are working at. New designers often have problems in keeping this straight. If their game components (size of the map, number of playing pieces) are not adequate to the level of detail they have in mind, they're going to have big problems, and the game is not going to work. You must, in a case such as that, either change the size of your game or change your level of abstraction.

When putting together your game prototype, you should think back on how you react when encountering a new game for the first time. A lot of psychology goes into designing these games. We are taking advantage of a lot of quirks in the way people think. All games do that, and the general game system presented here is the end result of over a thousand games designed in the past and millions of game-playing sessions. I don't presume to know exactly what goes on in people's minds when they play a game, but I know what works.

One of the things that you have to wrestle with to make your game work are the victory conditions. These are the objectives both sides are pursuing in order to win the game. These should also be the goals of the historical conflict the game is based on. Sometimes, the historical victory conditions are straightforward, but not always. Wargames based on battles that are part of larger campaigns are often more complicated situations. A prime example is the *Drive on Metz* game in this book. Both sides would consider it a victory if they could simply destroy all of the other side's combat units. But this was very unlikely and it is almost always very unlikely. Even in older periods, where the historical account speaks of one side "destroying the enemy host," most of the enemy host simply ran away. In *Drive on Metz,* seizing the city of Metz or simply getting by it is the primary U.S. victory condition. For the Germans, victory lies in preventing an American victory and doing it with as few troops as possible. The Germans had many other battles to fight at that time, and they needed all the soldiers they could muster. So don't consider victory conditions simple or easy: They're not. But coming up with workable and realistic victory conditions is often one of the most stimulating and rewarding mental exercises you'll encounter in designing a game.

You are well advised to begin by designing a fairly simple game. This will hone the basic skills you will require to handle any game-design chore, and will do it in a way that will not leave you frustrated from tackling something beyond your abilities. Even the largest and most complicated game is basically an application of the techniques I am detailing here. It is simply a question of being able to have more things flying through the air at the same time.

Getting back to our 10 steps, we find that we are now up to step 4. We have a prototype of the game. We have a map. We have playing pieces. We have a Combat Results Table and Terrain Effects Chart. The next step is to write the rules.

Beginning on page 132 are the game's rules. These were prepared by me in the space of a few hours. I used what we called at SPI a ''rules master.'' It is a basic set of rules for no game in particular and was stored in one of the computers. All the staffer has to do is sit down at the terminal and order the machine to change all references to ''First Player'' to ''American'' and all references to ''Second Player'' to ''German'' and make a few other minor changes. One must type in one's Terrain Effects Chart, Combat Results Table, order of appearance, victory conditions, and the like. It takes a few hours and provides one with a first draft of the rules. This is a common procedure learned the hard way. In fact, I was first introduced to this procedure by Tom Shaw of Avalon Hill years ago when he asked me to design my first game for them, on the naval battle of Jutland (in 1916). I asked, ''How do I start writing the rules?'' He said, ''It's simple. You simply take the last game we published and use it as a model.'' I thought for a moment and then spoke up: ''The last game was *Guadalcanal*'' (a game of U.S. Marines and Japanese infantry slugging it out on a tropical island in the South Pacific in 1942). Aside from that apparent mismatch, the idea is basically valid. It's not as easy to do as it used to be because of the great proliferation of different game systems, but you can usually find a game that matches the one you are currently working on. With the current proliferation of PC-based text scanners, I know of more than one game designer who has scanned another game's rules, done some heavy editing, and finished his own rules-writing job much more quickly than in the pre-scanner days.

A good example of the ''rules master'' approach occurred when, in mid-1979, I designed and developed a game called *Demons: The Game of Evil Spirits*. This was a fantasy-type game based upon the Lesser Key of Solomon, a medieval magical system for finding treasure. As the game system developed, it progressively became a solitaire game. Although it was a rather simple game, I was using a rules master for two-player games. This required a considerable amount of editing on my part and the critical contribution of half a dozen other people (testers and other staff) to get all the wrinkles out of it. When I finished, though, I had a good rules master for small solitaire games that I promptly appropriated before it was even published (I grabbed the galleys) for a game I did a few months later called *TimeTripper* (the classic case of the 20th-century soldier getting bounced around through a bunch of historical battles as he shoots his way back home). This one also worked very well as a solitaire game, and—what do you know—there was a rules master there, even if I did have to make it myself.

Step number 6, as I mentioned earlier, is the most difficult one. This is game development: taking all these neatly put-together little items and exposing them to the harsh criticism of the player. Play-testing usually isn't too difficult. There are plenty of gamers around who are more than willing to sit down and play a brand-new game. Most play testers are also keenly aware of the fact that they are in a position to make a substantial contribution to the game's final quality. While the designer and the developer are the people who get their names on the cover of the game, it is not for nothing that the testers are listed under the more extensive credits found inside the game itself.

The biggest single problem one has with play testers is when you get one who would rather play the game than test it. When I began testing most of my games, I didn't really have any victory conditions other than the obvious ones that are apparent in the game situation, particularly if it's a historical situation. The players generally know something about the situation. They know who won originally and some of the reasons why. Of course, more precise victory conditions must go into the game eventually, but in the early stages you are merely testing the mechanics of the game. If someone at this point merely plays to win, he is going to have two problems. One, he doesn't have a precise definition of what winning is, and, two, he is going to take advantage of obvious loopholes in the rules as they exist at the beginning of testing. Quite often, this is merely a lack of maturity. I have used testers as young as 12 or 13 years of age. These fellows are usually quite precocious, and as long as they kept their minds on what the testing was supposed to accomplish (the mechanical debugging of the game), I was glad to have them. With older testers, this is much less of a problem. The best play tester you have available to you is yourself. People are often fearful of their own creations, but these fears are generally baseless. When it comes to playing the game, you simply sit down and put yourself into "play mode" and play the damn thing. Turning it over to someone else for play-testing is something of an acid test and a double check. But you'd be surprised, when you think about it, by how many things you caught yourself and how few had to be caught by a play tester. Thus, you might overlook the fact that the smallest error in the rules or the game system can become a critical impediment to the game's working. I say this to drive home the point that you can catch a great deal but not all of the problems in a game yourself, by playing it by yourself before having to go to play testers. This is important also because even when using testers, there's always a communications problem. What they may see as a problem may simply be a result of their playing with the wrong rules. Again, this is the reason for getting your rules down on paper as soon as possible. When you're playing the game yourself, you still have the problem of misinterpreting your own rules, but you're more aware of the problem, and it's much simpler to solve it, unless you're prone to losing arguments with yourself.

During this testing process, you make changes in the rules as necessary. Depending upon the complexity of the game, you might have to play it anywhere from a half a dozen times to dozens of times during the testing stage. This is only a fraction of the testing the game will get before it's finished. The bulk of this testing will take place during the blind-testing phase.

Blind-testing is nothing more than having the game played without the designer or developer in attendance. You are, in effect, sending out the closest thing to a finished version of the game, and the people who are playing it are approaching the game as if they had just bought it in a store. After a few rounds of blind-testing (occasional in-house testing will also be continued to check various ''patches'' in the game), your game will be finished.

I am assuming that many of you could follow all of these procedures. All you need is a few game-playing friends to help you with your play-testing, and some other people, either local or otherwise, to help you with your blind-testing. All of the corrections and patches on the rules can be made by simply editing the manuscript on your PC and printing out a new set. It's also useful to put the date of the change in brackets (as in [3 Jun 93]) next to each change, so you, and the testers, know what was changed and when.

The eighth step is one you really don't have to go through unless you are dying to get the game published. This is the editing, wherein somebody else takes your finished product and tries to pick holes in it. This should be someone with some experience in dealing with game rules, because he will be looking for very subtle things. Most people who work in publishing in a copyediting function (not just copy editors, but most editors in general) could perform this function. It isn't really all that necessary if the game is simply for your own and your friends' amusement and entertainment. However, if you want to get the game published, send it to a publisher, and, if it meets the standards set forth in this section and it's on a salable subject, it stands a good chance of being published. Many of the smaller wargame companies are interested in outside submissions. Your chances of getting your game published are improved tremendously if you submit it to the publisher in the format I have just outlined.

The ninth step is handled by a publisher, whether that is yourself, if you're actually going to prepare a limited edition of the game for distribution, or a regular publisher. Every time the copy is retyped, primarily for typesetting, and then pasted up for printing, there is a probability of something getting scrambled or dropped or whatever. At this stage of the game, you simply have to sit down and reread the copy quite a lot.

The last step is the one that never ends. This is the evaluation of the game. Was it good? Was it bad? For what reason? The results here are often ambiguous because any game, no matter how "bad" it is considered to be by the majority of people, has a small but dedicated group of admirers. I have found this to be true even for games that not only I, but also a majority of gamers, disliked. When talking to the small group of admirers, I find that there were indeed elements in this game that did have merit. It's simply a question of what you're looking for and how you look at it. Frankly, I like nothing better than to sit around and mouth off about what game was good and what game was bad. In fact, parts of this book consist of quite a lot of that.

## WHY THE RULES ARE THE WAY THEY ARE ✱✱

The rules to *The Drive on Metz* follow a rather strict format. This is done for a reason—in fact, for a number of very important reasons.

The most important reason for having formats for the rules is that it makes it easier for the gamer to, first, learn the game and, second, to refer to a particular aspect of the game rules while playing the game.

The current system of writing game rules was developed by Redmond Simonsen (my partner at SPI) in 1970. The system underwent a fairly major revision in 1979 and saw more revisions during the 1980s, especially by Bob Ryer and Mark Herman at Victory Games. My personal approach to game rules deviates somewhat from the "standard," and the same can probably be said for just about anyone who actually writes game rules. For example, I tend to divide game rules into three parts. The first part would be sections 1 through 3, which would include all of the introductory and "housekeeping" material up to and including section 3.0, which would be the basic procedure or the sequence of play. The second part of the rules would be merely an elaboration on the sequence of play, which in many cases it is. The third section of the rules would be optional rules and scenarios.

Another modification I advocate is putting the initial set-up (at least one of the scenarios) in the front of the rules, along with the victory conditions. This is because I feel most gamers, especially experienced ones, want to get right into it, at least to set it up, and pretty much learn the rules by bouncing around inside the game.

In order to do this, the gamer really only needs the first third of the rules—that is, the introduction, the inventory of the game parts, glossary of terms (which should include any special game terms), the scale (which is as important as historical information), the description of the game map and the playing pieces, and then the general set of instructions for the initial scenario. Something else I prefer to see is what I call the "summary sheet." This is one piece of paper (whether 8½ × 11 or 11 × 17) that contains the essential elements necessary to play the game—that is, the Sequence of Play and most of the charts with as little explanation as possible, only such explanation as is necessary to constantly clarify a complex procedure.

This system of writing rules became known as "the case system," in which each of the major rules is initially stated rather briefly and in general terms. Then it is described in more detail, and finally a series of "cases" is given. These cases are usually one- or two-sentence affairs, each describing a specific element of the rule. Take, for example, the rule in *The Drive on Metz* on the movement of units. The general rule simply states that each unit has a movement allowance number printed on it that represents the basic number of hexes a unit may move in a single Movement Phase. Each player moves only his own units during the Movement Phase of his Player Turn (as outlined in the sequence of play).

The procedure gets into more detail:

> Units move one at a time, hex by hex, in any direction or combination of directions the player desires. The Movement Phase ends when the player announces he has moved all of his units that he chooses to (or at the time he begins to make any attacks).

The cases then somewhat belabor the obvious. For example, case 4.1 simply says a unit may never exceed its movement allowance and then a paragraph is devoted to elaborating on that point. Case 4.2 says units must spend more than one movement point to traverse some types of terrain.

Ideally, the case system is supposed to provide the experienced player with sufficient reminders in the general rule and procedure to allow him to play the game without going through the entire rule. For people learning the game, who are having a problem on some minor technical point, the cases then provide the necessary elaboration.

Work is constantly going on at many of the game publishers to develop more efficient ways of getting the rules across to the players with the least amount of effort on the gamers' part. Unfortunately, the people working on the games will often take the easy way out and do things the same old way where additional effort might have produced a more efficient rule. Writing rules is not one of the more "glamorous" aspects of working on games, a task that is in general more drudgery than glory.

The rules themselves represent the biggest dollar and time investment in any game. The playing pieces and the game map, even the box cover, may look much more colorful and interesting. But the rules are where the money is, or at least they are where the money should be. Without good rules, you cannot have a good game.

**The Drive on Metz: September 1944**





**[1.0] INTRODUCTION**

*The Drive on Metz* is a two-player game re-creating General George Patton's attempt to seize the key city of Metz and get across the Moselle River before the retreating Germans could form an effective defense. The American forces (three divisions of the XX Corps) had just completed an epic pursuit across northern France after the Allied breakout from the Normandy beachhead. Allied forces were at the end of an exhausted supply line and had barely enough resources for one last push. If they had been able to get across the Moselle River they would have compromised the entire German Westwall defenses. This would have enabled Patton to make an attempt at crossing the Rhine before the end of 1944 and possibly ending the war months earlier. The defending German forces, four divisions of the 82nd Korps of the German 1st Army, where a combination of hastily collected and organized units, including fresh units from the interior and remnants of units that Patton had pursued across France from Normandy. The battle, which began on 7 September and lasted for about a week, was a one-time chance that could have gone either way. Historically, the Germans won. But just barely.

The game has a slight bias to the German player, to reflect the reality of the historical situation. It is a real test of your decision-making ability. Time pressure is on both players. The game is very unforgiving of mistakes. Both sides must never forget the victory conditions. One moment's inattention is sure to bring disaster.

After playing the game a few times, try adding the optional rules in Section 8.0. These rules add more balance to the game and are an accurate reflection of choices available to commanders in battle.

### [1.1] INVENTORY OF GAME PARTS

All of the game components are contained in this book. They consist of:

A game map (page 18)
Playing pieces (page 16)
Game rules (pages 132–143)
Game charts and tables (page 144)

Also needed to play the game is a six-sided die, which you will have to supply yourself. Lacking a die, you can use six pieces of paper numbered 1 through 6 and drawn from a container. You're better off with a die.

### [1.2] GLOSSARY OF GAME TERMS

There are no special terms used in this game that are not found in the "Technical Terms of Wargaming" section of this book (pages 66–86).

### [1.3] THE GAME SCALE

The game scale for the map is 1:250,000 (one hexagon = 4 km across). Each game turn represents one day of real time.

### [1.4] THE GAME MAP

The game map represents the area of Lorraine (France) just east of the Moselle River in the vicinity of the ancient fortified city of Metz. The area of the game map is 44 × 36 km. A hexagonal grid has been superimposed to regularize movement and the position of units.

**[1.5] THE PLAYING PIECES**

The playing pieces represent the combat units that actually took part in the original battle. Below is a description of the symbols found on the playing pieces:

| | | |
|---|---|---|
| | III | UNIT SIZE [III] SYMBOL [REGIMENT] |
| UNIT DIVISIONAL IDENTIFICATION [5TH] | 5 ⊠ 11 | UNIT REGIMENTAL IDENTIFICATION [11TH REGIMENT] |
| COMBAT STRENGTH [5] | 5–4 | MOVEMENT ALLOWANCE [4] |

The unit-type symbol is that of an infantry unit.

**[2.0] SETTING UP THE GAME**

To actually play *The Drive on Metz,* you will have to make a photocopy of or remove page 18. The playing pieces are another matter. You could tear out the page with the playing pieces (page 16) or photocopy that page. You must then cut out the pieces. Normally, they are printed on thick die-cut cardboard.

So if you have a wargame or two, just scrounge together twenty blank counters and make your own counters for *The Drive on Metz*. Another good idea is to paste the page with the playing pieces to a piece of cardboard. Then cut the pieces out. All the other game components are contained in the book (page 144).

**[2.1] GENERAL COURSE OF PLAY**

The German player sets up his units on the map. Then the American player and the German player move alternately for seven turns. At that point, you consult the victory conditions to determine who, if anyone, has won the game.

**[2.2] LAYING OUT THE GAME COMPONENTS**

All but one of the German units are placed on the game map at the beginning of the game. All American units enter the map from the west edge of the map as listed in Rule 7.0. Lay the map flat on a hard surface, using tape if you like, to keep it in place.

**[2.3] WHO SETS UP FIRST**

The German player sets up the following units first (units are identified as follows: division ID/regiment ID/combat strength movement allowance/ set-up hex):

559 / 1125 1-4/0502
559 / 1126 1-4/0403
462 / UTRFHR 2-4/0505 Unterführer (NCO school staff and students)
462 / 1010 1-4/0507
462 / FHNJKR 3-4/0509 Fahnenjunker (Officer Candidate School)
3PG / 8PG 2-8/0609 PG—Panzergrenadier (motorized infantry)
3PG / 29PG 2-8/0611
17SS / 3855 2-8/0709 SS = Schutzstaffel (Nazi party troops)
17SS / 3755 3-8/0808
Metz / 0807-Metz Garrison

See rule 7.0 for American Units and German Reinforcements.

**[2.4] HOW TO WIN**

Players win by obtaining more victory points than their opponent. Victory points are obtained as follows:

**For the Americans:**

Five points for each unit on the east side of the Moselle for three complete and continuous game turns before the end of the game.

Five points for an American unit being the last unit to enter or pass through Thionville (hex 0701).

Twenty points for an American unit being the last unit to enter or pass through Metz (hex 0807).

Five points for each unit to exit the east side of the map before the end of the game.

**For the Germans:**

Ten points for each unit to exit the west edge of the map by the end of the game.

Seven points are obtained for each unit of the 3rd Panzergrenadier or 17th SS Panzergrenadier Division to exit the east or south edge of the map by the end of the game. The number of victory points varies according to the game turn in which the unit is exited. If the unit exits on Turn one, 7 points are obtained. Turn two, 6 points; Turn three, 5 points; Turn four, 4 points; Turn five, 3 points; Turn six, 2 points; Turn seven, 1 point.

Victory is determined by adding up each side's points at the end of the game and comparing the totals. If one side has five more points than the other, it has won a marginal victory. Ten points is a substantial victory and 15 or more points is a decisive victory.

The victory conditions represent the goals of the two sides. The Americans wanted to get across, and beyond, if possible, the Moselle. In addition, the Americans wanted to seize the two key cities in the area to deny the Germans access to their key locations as transportation centers. The Germans wanted to prevent the Americans from achieving their goals while at the same time getting their valuable mobile units out of the battle so they could be used as reserves and/or be rebuilt. The Germans would also seize any opportunity to get into the American rear area. German doctrine, even on the defense, was very aggressive.

### [2.5] PLAYER'S NOTES

*American.* Since the game has a slight bias toward the Germans, the Americans must work very hard to gain a victory. The Germans can gain victory points quickly at the beginning. These must be offset and victory points built up. There are three major routes of advance for the Americans: toward Thionville in the north, toward Metz in the heavily defended center, and across the Moselle in the south. Don't make the mistake of concentrating on one axis of advance. That's too easy for the Germans to block. Go where the victory points are. Keep in mind the "Principles of War": maintenance of the objective, concentration of force, economy of force and security. It is easy for the American to become involved in side issues. Don't allow that to happen. There isn't time for anything that doesn't move units directly toward the main objectives.

*German.* Careful planning is the key to German victory. Holding the Americans off is not enough. You have to build up victory points quickly by moving units off the map. Don't be misled by the victory conditions. Moving too many units off early will build up victory points. However, that can make it impossible to hold off American assaults. Note that you want weaker units to move off the map. Victory points are given for when units move off, not for their size. Tactically, terrain will be your best friend and worst enemy. Avoid being trapped with your back to a river. That's one of the easiest ways for units to be eliminated.

### [3.0] BASIC PROCEDURE

#### The sequence of play

The players take turns moving their units and making attacks. The order in which they take these actions is described in this sequence of play outline. One completion of the sequence of play is called a game turn. Each game turn consists of two Player Turns. Each Player Turn consists of two phases.

**The American Player Turn:**
**Phase one: the American movement phase**
The American player may move his units. He may move as many or as few as he wishes, one after another, within the limitations of the rules of movement.

**Phase two: the American combat phase**
The American player may attack adjacent enemy units. He may perform these attacks in any order he wishes, applying the results immediately as each attack is made.

**The German Player Turn:**
**Phase three: the German movement phase**
The German player may move his units. He may move as many or as few as he wishes, one after another, within the limitations of the rules of movement.

**Phase four: the German combat phase**
The German player may attack adjacent enemy units. He may perform these attacks in any order he wishes, applying the results immediately as each attack is made.

These two Player Turns are repeated six times. The game is then over and the players determine the victor.

### [4.0] THE MOVEMENT OF UNITS

**General rule:**
Each unit has a movement allowance number printed on it which represents the basic number of hexes it may move in a single movement phase. Each player moves only his own units during the movement phase of his Player Turn, as outlined in the sequence of play.

**Procedure:**
Units move one at a time, hex by hex, in any direction or combination of directions that the player desires. The movement phase ends when the player announces that he has moved all of his unit that he chooses to.

**Cases:**

**[4.1] A unit may never exceed its movement allowance.**
During its movement phase, each unit may move as far as its movement allowance permits. Basically, each unit spends one or more movement

points of its total allowance for each hex that it enters. Individual units may move less than their movement allowance. Units are never forced to move during their movement phase. Units may not, however, lend or accumulate unused movement points.

**[4.2] Units must spend more than one movement point to traverse some terrain types.**
The basic cost to enter a clear terrain hex is two movement points. The basic entry cost to enter some terrain hexes, however, is higher. These costs are specified in the Terrain Effects Chart. If a unit does not have sufficient movement points to enter a given hex, it may not do so.

A hex containing more than one type of traversable terrain is entered at the higher of the two costs.

When a unit enters a hex through a road hex side, it pays only the cost for moving one hex along the road, regardless of the type of terrain entered. Conversely, a road has absolutely no effect on movement if the hex is entered through a nonroad hexside.

**[4.3] A unit may never enter or pass through a hex containing an enemy unit.**

**[4.4] A unit may never end its movement phase in the same hex as another friendly unit.**
One or more units may move through a hex containing another friendly unit, but the moving units may never end the movement phase in the same hex as another unit. If this should inadvertently happen, the opposing player gets to choose which of the illegally placed units are to be destroyed (so that only one remains in the hex).

**[4.5] A unit must stop upon entering a hex adjacent to an enemy unit.**
Whenever a unit enters a hex that is directly adjacent to any of the enemy player's units, the moving unit must immediately stop and move no farther. Note that there are six hexes adjacent to most hexes on the map. The six hexes adjacent to an enemy unit are called the Zone of Control of that unit.

If a unit begins the movement phase of its turn adjacent to an enemy unit (i.e., in its Zone of Control) it may not leave that hex except as a result of combat (either the enemy unit is destroyed or retreated or you are retreated).

**[4.6] Units may not leave the map as a result of combat.**
If forced to do so by the Combat Results Table, they are eliminated instead. Units may leave the map to obtain victory points. When they do leave they may not return.

## [5.0] COMBAT PRECONDITIONS

### Eligibility requirements for attacking units

**General rule:**
Each unit has a combat strength number printed on it which represents its basic power to attack and defend. During its combat phase each unit may participate in an attack against an adjacent enemy-occupied hex.

**Procedure:**
The player examines the positions of his units, determining which are adjacent to enemy units. These are the units that are eligible to conduct attacks during that combat phase. Attacks are conducted using the Combat Results Table and the procedures detailed in the section on combat resolution.

**Cases:**

**[5.1] A unit is never forced to attack. Attacking is a purely voluntary action.**
In a given combat phase, some of the eligible units may attack and others may not. Indeed, the player may totally pass up the chance to make any attacks at all during a given combat phase.

**[5.2] Only one enemy-occupied hex may be the object of a given attack.**
Even though an attacking unit may be adjacent to more than one enemy-occupied hex, it may conduct an attack against only one such hex in its combat phase.

**[5.3] No unit may participate in more than one attack per combat phase.**

**[5.4] No unit may be the object of more than one attack per combat phase.**
Regardless of how many attacking units are adjacent to it, a given enemy unit may only be subjected to one attack per combat phase. It must defend against this attack; unlike the attacker, the defender's participation is involuntary.

**[5.5] More than one unit may participate in a given attack.**
As many units as are adjacent to an enemy-occupied hex may combine their strengths into one attack against that hex. Remember, however, that

if one or more such units attack, this does not obligate any of the other adjacent units to participate.

## [6.0] COMBAT RESOLUTION

### How attacks are evaluated and resolved

**General rule:**
An "attack" consists of the comparison of the strength of a specific attacking force with that of a specific defending force resolved by the throw of the die in connection with a Combat Results Table. The results may affect either the attacker or the defender.

**Procedure:**
The attacking player totals up the combat strength of all his units that are involved in a given attack and subtracts from that total the combat strength of the enemy unit being attacked. The resulting number is called the combat differential. The player locates the column heading on the Combat Results Table that corresponds to the combat differential. He then consults the Terrain Effects Chart to see if the column of combat resolution is to be shifted because of the terrain the defending unit is on. If more than one type of terrain exists on that hex only the worst (for the attacker) is used.

He rolls the die and cross-indexes the die number with the combat differential column and reads the result. The indicated result is applied immediately, before going on to any other attacks. When he has made all of his attacks, the player announces the end of his combat phase.

**Cases:**

**[6.1] The attacking player must announce which of his units are involved in a given attack against a specific defending unit.**
He must calculate and announce the combat differential, specifying which of his units are participating in the attack before it is resolved. He may resolve attacks in any order he chooses. Once the die is thrown, he may not change his mind.

**[6.2] The calculated combat differential is always determined to represent a specific column of results on the Combat Results Table.**
If the combat differential in an attack is higher (or lower) than the highest (or lowest) shown on the table, it is simply treated as the highest (or lowest) column available.

**[6.3] The abbreviations on the Combat Results Table will indicate which units are retreated.**

*Ar:* Attacker retreats; all the units involved in the attack are forced to move one hex away from the defender. Defending unit has the option to advance after combat.

*Dr:* Defender retreats; the defending unit is forced to move one hex away from the attacking unit(s). One of the attacking units may advance after combat.

*Dr2:* Same as dr except the defending unit must retreat two hexes (see 6.4).

**[6.4] Movement as a result of combat**

When a unit is retreated, it may retreat only if it does not have to enter a hex containing another unit (enemy or friendly) or enter a hex adjacent to an enemy unit (enemy Zone of Control hex). Retreating units may not cross rivers when retreating. Any unit that cannot retreat because of the above is destroyed and removed from play.

Whenever a unit vacates a hex as a result of combat, one of the victorious units may enter the vacated hex(es).

**[6.5] Effects of terrain on combat**

The Terrain Effects Chart shows how many columns the combat is shifted when the defender is on certain types of terrain. If more than one type of terrain exists on a hex only the worst (for the attacker) is used. The effects of more than one type of terrain (like a fortress in the woods) are not cumulative.

## [7.0] REINFORCEMENTS

### How additional units enter the game

**General rule:**

In addition to the force with which they start the game, both players receive units during the movement phases of specified game turns (see the schedule of reinforcements on pages 143–144).

**Procedure:**

At any time during the specified movement phase, newly arriving units may enter the map in the hexes indicated.

**Cases:**

**[7.1] When reinforcements arrive on the map, they behave identically to units already on the map.**
The arrival (into the proper hex) costs the reinforcing units the appropriate number of movement points for that terrain type. If entering on a road, it is assumed that they are entering the map through a road hex side. The units move (and they may participate in combat) in the Player Turn of arrival.

**[7.2] If the entry into the arrival hex cannot be performed as a legal move, the reinforcing units may be brought in at the closest hex at which it would be legal to place them.**
If, for example, the arrival hex is enemy-occupied, the reinforcing units would be diverted to the closest hexes not occupied by enemy units. If possible, however, units must enter in the hexes specified. Note that if the entry hex were enemy controlled, only one unit could enter there (and would stop in that hex).

**[7.3] The entry of reinforcements may be delayed for as long as the player wishes.**
Should the player so desire, he may hold back all or part of the reinforcements due him in any game turn. He should keep a record of any such delayed reinforcements. He need not reschedule their appearance; they may be brought in at will in any of his subsequent movement phases.

**[7.4] American reinforcement schedule**
The following American units enter the game map on Turn one. They are identified as follows: division ID/regiment ID combat strength-movement allowance entry hexes (which hexes units may enter map on).

90 / 358 4-4/0101-0104
90 / 357 4-4/0101-0104
7A / CCA 7-10/0105-0107
5 / 2 5-4/0105-0107
7A / CCR 5-10/0108-0111
7A / CCB 7-10/0108-0111
5 / 11 5-4/0108-0111
5 / 10 5-4/0108-0111

A-Armored, CCA-Combat Command A

### [7.5] German Reinforcement Schedule

The following German units enter the game on the turns indicated (see 7.4 for how to read the listing).

Turn two

106PzB 1-8 / 0401-0901 PzB-Panzer [tank] Brigade

### [8.0] OPTIONAL RULES

The basic game is biased somewhat toward a German marginal victory in order to reproduce the result of the historical campaign. You can balance the game by agreeing to use one or both of the following optional rules before the game starts. Try these after you feel comfortable with the basic game.

### [8.1] COMMAND CONTROL

This optional rule will usually help the Americans. It stipulates that if in a given attack the attacking forces contain more than one unit from the same division, the attacker receives a one-column shift to the right on the Combat Results Table. This rule reflects the benefit of having units that are used to working with each other fight together, and from having higher-level commanders present.

### [8.2] TACTICAL WITHDRAWAL

This rule usually helps the Germans, although it can be very useful to the Americans. The rule is: Armored units (units with a movement allowance of eight or more) may, in their movement phase, withdraw from the Zone of Control of an enemy unit. They do this by paying four additional movement points (in addition to the normal terrain costs) to move to a hex not in the Zone of Control of an enemy unit. The first hex moved to may not be toward the enemy's edge of the map (east for the Germans, west for the Americans), nor may it be across a river. The unit exercising tactical withdrawal may continue to move after the first hex, if it has any movement points remaining.

### [8.3] OPTIONAL SCENARIO

Most games have a number of additional setups for the game. These are called scenarios and this is an example based on *The Drive on Metz.*

*Scenario:* Screaming Eagles over Lorraine. Patton convinces Eisenhower that a regiment of the 101st Airborne Division would be more useful dropped near Metz than used (as it historically was) in the Netherlands for Operation Market Garden. On any turn of the game, the American player receives the 502nd parachute regiment of the 102st Airborne division (a 3-4 unit). This unit may be placed on any clear terrain hex on the map.

## TERRAIN EFFECTS CHART

| *Terrain* | *Example Hex number* | *Effect on movement* [MP's to enter] | *Effect on combat* [Leftward column-shifts on CRT] |
|---|---|---|---|
| Clear | 0406 | 2 | None |
| Forest | 0404 | 4 | 2 |
| Rough | 0306 | 3 | 1 |
| Town | 0206 | Same as other terrain in hex | 2 |
| Fortified | 0507 | Same as other terrain in hex | 3 |
| Road | 0405 | 1 | None |
| River | 0804 | Must be adjacent at start of movement, uses all MP's to cross | 3 [Only if all attackers are attacking across] |

Note: Only one terrain effect may by used by a defender in combat.

## COMBAT RESULTS TABLE

| Die roll | Differential [attacker's strength minus defender's] | | | | | | | |
|---|---|---|---|---|---|---|---|---|
| | −1+ | 0 | +1 | +2, +3 | +4, +5 | +6, +7 | +8, +9 | +10+ |
| 1 | – | DR | DR | DR | DR2 | DR2 | DR2 | DR2 |
| 2 | – | – | DR | DR | DR | DR2 | DR2 | DR2 |
| 3 | AR | – | – | DR | DR | DR | DR2 | DR2 |
| 4 | AR | AR | AR | – | DR | DR | DR | DR2 |
| 5 | AR | AR | AR | AR | – | DR | DR | DR |
| 6 | AR | AR | AR | AR | AR | – | DR | DR |

**–: No result, DR: defender retreat one hex, AR: attacker retreat one hex, DR2: defender retreat two hexes.**

Credits:

Design and Development: James F. Dunnigan.
Play-testing: Richard Bartucchi, Gary Gillette, Dave Rodhe, Bill Watkins, and a few other folks whose names I forgot to write down.
Graphic Design Assistance and Inspiration: Redmond A. Simonsen.
Graphic Production: Ted Koller, Bob Ryer and the folks in the William Morrow Production Department.
Forbearance: Susan Hanger.

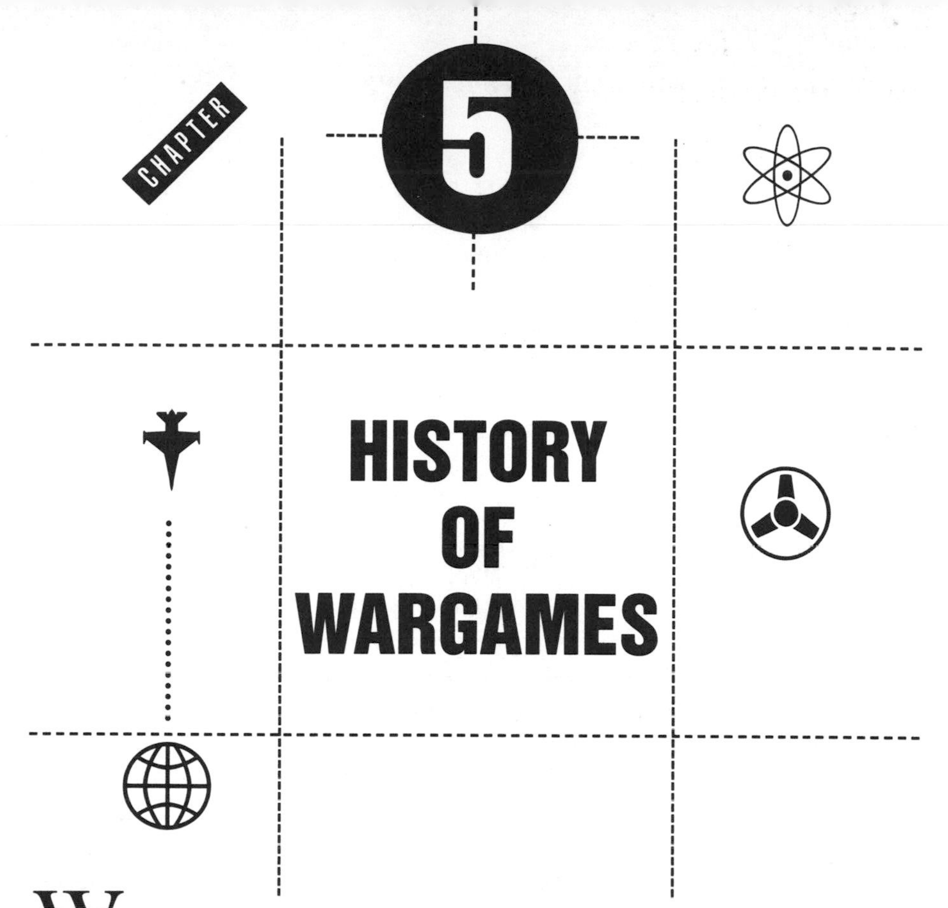


# HISTORY OF WARGAMES

Which came first, warfare or wargames? Given the lethal nature of actual warfare and man's penchant for self-preservation, it is quite possible that some form of wargame occurred before the first organized war. Whatever the case, wargames have been around for a long time. Warfare may have got more attention, but wargames are a lot safer.

Chess is one of the oldest surviving ancient wargames, and games similar to chess go back thousands of years. Chess is also one of the more accurate wargames for the period it covers (the pre-gunpowder period). Chess is a highly stylized game. It is always set up the same way; the playing pieces and the playing board are always the same. The board is quite simple. Each of the pieces has clearly defined capabilities and starting positions, much like soldiers in ancient warfare. Given that ancient armies were so unwieldy and communication so poor, it is easy to see why each player in chess is allowed to move only one piece per turn. Because the armies were so hard to control, the battles were generally fought on relatively flat, featureless ground. Then as now, the organization of the army represented the contemporary social classes. Thus, the similarity between chess pieces and the composition of ancient armies.

As a minor point on the history of chess, the "queen" was, until quite recently, called not the "queen" but the "general," "prime minister," or other, similar titles to represent the piece's true function, namely, the actual head of the army who had under his personal command the most powerful troops. This is why the "queen" piece is so powerful. Not only does it represent the single-best body of troops, but also the very leadership of the army. The king, on the other hand, is indeed the king of the kingdom, without whose presence the army is lost. Thus, the king is not necessarily a soldier of any particular talent. During the battle, his main function is to survive and to serve as a symbol, a rallying point for his army.

For thousands of years, chess and variants of chess were used by civilian and military personnel alike for entertainment, education, "simulation." As more education, leisure time, and technical sophistication became available, the games themselves expanded in a similar fashion. In the 17th century, the first modern wargames appeared, and within 200 years wargames surpassing the complexity of most (but not all) of the games covered by this book came into existence. These earliest wargames were simply elaborate variations on chess that replaced the traditional components of chess with playing boards that represented real terrain and playing pieces that accurately (to one degree or another) simulated contemporary troops and their capabilities. Many of these early efforts were prepared by civilians, as professional soldiers in that period were chosen more for their courage and loyalty than for any desire to invent new things. These civilian wargames were often lacking in crucial elements of reality, as their authors often had minimal military experience.

By the early 19th century, Prussians, civilians as well as members of the Prussian Army, developed the first detailed and realistic wargames. These were used for training, planning and testing military operations. The mechanics of the games were developed after careful study of actual military maneuvers and battles. After the wars of German unification concluded in 1871, the Germans made no secret of their new technique, and most European armies quickly followed their lead. However, no one took it as seriously as the Germans, or got as much out of it.

About the turn of the century, the famous science-fiction writer H. G. Wells wrote a book called *Little Wars*. This book described a somewhat simpler form of the wargames than those used by the professionals. Wells's game used toy metal soldiers to represent the military units, and is another of the direct antecedents of contemporary wargames.

Up until World War II, the majority of the wargames available involved single battles. The planning for larger operations was not so much a game as it was a paper-shuffling exercise directed toward solving the puzzle of getting all the pieces moving at the right place and time, much like planning a railroad schedule. But during World War II, things began to change.

Much of the gaming used in World War II was of the conventional sort. But equally, if not more, important was the introduction of more scientific techniques. Much of the "gaming" that took place at the behest of the military after World War II was more operations research (OR) and systems analysis than the study of history. (See Chapter 9 for more on this period.) The study of past military operations, and history in general, which had formed the basis of the earlier wargames, was very much neglected. This situation has only been rectified to any degree in the last ten years. Meanwhile, the primacy of OR in the military allowed civilian wargames to pull ahead of, and in many cases replace, functions previously performed by OR-based wargames. The military only began to play catch-up and develop effective games for its own requirements during the late 1970s and through the 1980s.

Civilian wargaming in the United States began in 1953, when a young gentleman from Baltimore named Charles S. Roberts developed a game called *Tactics*. It posited two hypothetical countries, with typical post–World War II armies, going to war with each other. The game was professionally produced and distributed through the Stackpole Company (which already had a reputation as a publisher of books on military affairs). This was the first of the modern commercial wargames (as we know them).

Charles Roberts was then working in the advertising business and was indulging in the commercialization of his hobby as a sideline. But by 1958, he realized that there were a lot of people who were interested in his type of game, and he founded the Avalon Hill Company. For the next five years, Avalon Hill experienced tremendous growth. But up until 1961, only six games were published. However, during 1961, an additional six games were published, and from 1962 to 1963 six more games were published. Of these eighteen, only nine were wargames. They included *Gettysburg, Tactics II, U-Boat, Chancellorsville, D-Day, Civil War, Waterloo, Bismarck,* and *Stalingrad.* The wargames, however, accounted for most of the sales, and by 1962, Avalon Hill was selling more than 200,000 games a year.

But then it all collapsed. There was a combination of problems. First of all, the distribution system for games was changing in the early 1960s. Many distributors were having a hard time, and a number of them, who represented 25 percent of Avalon Hill's volume, went bankrupt. Avalon Hill had borrowed heavily to finance its expansion, and this really left it on the ropes. Charles Roberts turned the company over to his two largest creditors and went on to a career in the printing industry. Tom Shaw, who had joined Charlie a few years earlier (they had been longtime friends), was the only member of the old Avalon Hill to stay on. Business was pretty bad through the end of '63 into early '64, but then Avalon Hill began publishing one or two games per year and also decided to publish a long-planned wargaming periodical called *The General*. This was a critical move, as it provided a forum for gamers to discuss subjects of common interest, and more important, to be aware that they were all part of a large group.

Eric Dott, the president of Monarch Printing, the largest creditor of the old Avalon Hill, was now making most of the decisions. He made the key decisions to keep the company going and showed how to keep it going. Dott eventually bought out the other creditor/owner and became the sole owner of Avalon Hill. In later years, Dott would step in as needed to keep things going, and this enabled Avalon Hill to continue as a presence in the wargaming market. Tom Shaw has also stayed with it, being the day-to-day manager of the company, and was largely responsible for dragging me into the business.

## HEY, LET'S START A WARGAME COMPANY IN THE BASEMENT! ✱✱

How I got into the wargame business is one of those odd, series-of-coincidental-events things that often turn out to have far-reaching consequences. The emergence of the wargame company (SPI) I founded in the late 1960s in effect signals the next major chapter in the history of wargaming.

I picked up on wargames in the early 1960s, when I was in the army. While I had always been interested in history, I had never been all that curious about military history. But while in the army, I came across some G.I.'s who played the original Avalon Hill games, and as I was in the military, it seemed a logical thing to get involved with wargames. When I got out of the army in 1964, I kept in touch with wargames in a casual way and became somewhat obsessed with the idea of using the games to teach, and better understand, history. Spending the next six years working my way through Columbia University gave me ample opportunity to do some writing on military history the way I thought it should be done. Note that at this stage I considered games a means of better understanding military history so I could write a better account of it. My goal of writing books took a detour when I got to know Tom Shaw down at Avalon Hill. We got along quite well, and in 1966 he asked me if I would like to design a game. At that point, I had no aspirations to design games professionally, but accepted the challenge anyway. A year later, Avalon Hill published my effort, a game called *Jutland,* based on the naval battle of the same name during World War I. A year later, in 1968, they published my second effort, *1914*, which covered the opening rounds of World War I.

After doing two games for Avalon Hill, and carefully observing how they did it, I decided that there had to be a more effective way to publish games. It was at that point I decided to found SPI (Simulations Publications, Inc.). This was done in a rather impromptu fashion, much like the old 1930s movies in which a group of bright young kids gather around and say, "Hey, gang, let's put on a Broadway musical in Dad's garage!"

I did pretty much the same thing in 1969. However, we were city rats, and our first venue was not Dad's garage but a windowless basement in New York City's Lower East Side district. Our neighbors in the basement were a puppeteer on one side and a pornographer on the other. A typical Lower East Side mixture, then and now. The initial staff was comprised of local gamers. I had come to know a number of other wargamers in New York City since 1964, and recruited as many as I could to form SPI's first staff. We had no money; so I borrowed a hundred dollars from Al Nofi, one of the original SPIers. I had to pay Al back in a month, but it was enough to get started. What we lacked in financial resources we made up for in a lot of energy and a few ideas. When SPI began, we had the basic concepts that have remained the cornerstone of what all historical wargame publishers are still trying to achieve. First of all, we wanted games published by gamers. This meant hobbyists controlling all of the game development, production, and marketing decisions. The second principle was one of publishing more games. At the time, Avalon Hill was only doing one or two a year, and Avalon Hill was the only show in town. The third principle involved being more directly responsive to gamer desires.

Initially, in early 1969, we were only thinking of designing and publishing games, advertising them through a new magazine called *Strategy and Tactics* (*S&T*), which had begun publishing regularly in 1967. *S&T* was the brainchild of Chris Wagner, who was at the time an air-force sergeant stationed in Japan. In the middle of 1969, Chris Wagner's *S&T* went bust. He was not able to get much beyond 1,000 subscribers, and that was not enough to make it a viable operation. Nobody else seemed interested in taking it over, and as we were planning on using it as our chief means of promoting our new line of games (Avalon Hill allowed no other advertising in its magazine), I had no choice but to become the new publisher of *Strategy and Tactics*. Thus, we found ourselves in the magazine-publishing business, in addition to our efforts to publish more games. Doing the magazine also brought graphic-design ace Redmond Simonsen into SPI. I knew that the magazine, and the games, needed a professional look. Simonsen was a native New Yorker and a wargamer, in addition to being a highly talented artist. So I made him an offer he couldn't refuse: half the business (we later shared some of this with some of the original staff). And together we proceeded to do the deed.

Those early days were pretty hairy. Anything was considered possible, and with that attitude we made things happen. In that period, many of the still-current concepts of designing and producing games were either invented or given some solid form. The process by which a game goes from concept to finished product was worked out because the pace at which we were working had to be highly organized or nothing would get done. When we realized that we would have to simultaneously get our first issue of *S&T* out and have our first six games ready to go before subscribers lost interest, we had to innovate and hustle. For a few frantic weeks in the summer of 1969, we called ourselves "the game-of-the-week club" because I did, literally design three games in three weeks. At the same time, I was teaching the other lads the fine points of debugging the games and writing the rules up in a consistent and legible format. There were other complications, as I had a full-time job (plus a part-time job), and was going into my last year as an undergraduate at Columbia (on an honors program, no less, demanding that a thesis be written). One of our key people, Al Nofi, decided to sail over to Spain and back during that summer. He was late getting back, as sailing back he encountered a storm, the sailboat was wrecked, and he was literally lost at sea for several weeks. He showed up, rather more sunburned and lean than normal, for the furious last few weeks of preparations for our debut of *S&T* with its new format (with a complete game in it). We had everything ready on the Labor Day weekend of 1969, and rather than wait for the post office to open on Tuesday, we stuffed every mailbox on Avenue C with the 1,000 issues of *S&T*.

The response from the readers was quick, and overwhelming. Many ordered all six of our new games, renewed their subscriptions, and provided us with enough cash to get the operation off the ground. I also aced the honors thesis. While I didn't get to Woodstock (I was tending bar, a second job, down the road from all the traffic jams that weekend), 1969 was an interesting year.

After struggling for two years just to get the thing off the ground, I developed a marketing/advertising campaign in 1971 that really got us rolling. Within two years, we were reaching more than half of the active gamers in the country. While in 1969, fewer than 100,000 wargames were sold, almost all by Avalon Hill, three years later, this number had more than doubled, largely because of our efforts. Ten years later, the number of wargames being sold was more than two million. Unlike Avalon Hill, we actively promoted other publishers' games. Thus, our promotional efforts gave all of the new publishers a leg up and expanded the reach of wargames even more. To give you an idea of what this promotion campaign meant, consider the history of wargame sales.

**MANUAL WARGAME SALES: 1960–1991**

**Since 1960, over 20 million paper (or manual) historical wargames have been sold. Unit sales of historical wargames (paper, or "manual" type games) per year indicated.**

| *Year* | *Wargames Sold* |
|---|---|
| 1964 | 62,000 |
| 1965 | 65,000 |
| 1970 | 129,000 |
| 1975 | 743,000 |
| 1980 | 2,200,000 |
| 1985 | 900,000 |
| 1990 | 450,000 |
| 1991 | 400,000 |

These are sales of historical wargames, excluding science-fiction and fantasy titles. Nearly 10 percent of these were games published in *Strategy and Tactics* magazine. As with most books, about half of all sales were concentrated in less than 50 of the best-selling wargames. Several of these paper wargames have achieved extraordinary sales figures. *PanzerBlitz* has sold over 300,000 copies, but it has been in print since 1970. *Squad Leader* has sold over 100,000 copies of the basic game, plus many more of the add-on modules. About 1,200 game titles were published during this period, most selling at least a few thousand copies. Avalon Hill, with the widest distribution, could usually rely on a decent game selling at least 25,000 to 50,000 copies. This number has come down a bit since the boom times of the late 1970s and early 1980s. But even today, Avalon Hill can move at least 25,000 copies of a decent game over its two-to-five-year publishing life. Smaller companies, with more limited distribution, can usually move at least a few thousand copies. SPI stood somewhere in the middle, being able to sell at least 5,000 of a title and moving over 30,000 copies of best-sellers. The games in *Strategy and Tactics* sold as many games as there were buyers of the magazine, plus a few thousand additional copies when some magazine gamc was later published separately.

Computer wargames did not enter the market until 1980. In that year, only about 100,000 computer units were sold. But by 1985, computer wargame sales had moved past manual wargame sales, and by the late 1980s, some individual computer wargame titles had sold more than

250,000 units after several years on the market. Currently, several million computer wargames are sold each year, up to 10 percent of the 25 million computer games of all types sold annually (not counting the Nintendo games). And this excludes all the fantasy titles that feature a lot of combat activity. Computer-game sales more than tripled between 1985 and 1992. While there are only about 30 million PCs (of all types) in homes, nearly as many are in commercial locations (where there are more games being played than management will generally admit). Surveys indicate that about two thirds of the homes of actual or potential wargamers contain PCs. Currently, several computer wargames each year exceed 100,000 units sold, though the average computer wargame sells more like 20,000–30,000 units.

During the first three years at SPI (1969–72), Redmond Simonsen further refined the standards for editing and designing game components. Simonsen also had a flair for editing, and this, combined with his artistic skills, created a system for presenting games that has never been surpassed and is still widely imitated. While I was self-taught in the wargame business, Simonsen had graduated from Cooper Union with a degree in design. Cooper Union is one of those uniquely New York institutions, with a huge endowment and no tuition. However, entry is competitive, and other graduates have told me that getting in was worse than any job interview they subsequently had to go through.

By providing a model, SPI spawned dozens of other game companies, each following the SPI system to one degree or another. This system emphasized keeping the cost down, and not having a mounted (paper map glued to cardboard) map board. The mounted map board was a habit Avalon Hill picked up from the mainline toy and game publishers, and increased the cost of the game considerably. Initially, SPI marketed games solely by direct mail rather than trying to get them into stores. Some new companies, such as Simulations Design Corporation (SDC) in California, even attempted to imitate SPI's idea of putting out a magazine with a game in it. The head of that firm, Dana Lombardy, came to us for advice in the early 1970s, which (as was our custom) we freely gave. SDC eventually folded, as did many of the other young publishers, but a large number survived. Although SPI published nearly 400 games between 1969 and 1982, the dozens of other smaller publishers have published that many and more (depending upon how you define "publishing") to date. Some of these smaller publishers developed highly innovative ideas and have themselves contributed to profound changes in the hobby.

The most innovative and influential of these new game systems was the role-playing game (*Dungeons and Dragons*) developed by Gary Gygax and Dave Arneson in 1973. The closest SPI ever came to this was a game we published in 1973 called *Sniper,* which involved man-to-man combat in an urban area. Players had a tendency to individualize their playing pieces in *Sniper.* But I, as the designer, did not bother to take it as far as *Dungeons and Dragons,* which was also the first, or at least the most widely successful, of the fantasy games.

SPI's first science-fiction game, *Star Force,* was published in 1974 and went on to become one of SPI's best-sellers. After *Star Force,* SPI published many other fantasy and science-fiction games, almost all of which did very well, and by the late 1970s, many of the smaller publishers realized that the quickest way to survival and success was to concentrate on fantasy and science-fiction topics.

This shift toward fantasy and science fiction somewhat dismayed many of the older gamers, the grognards. Most of the original wargamers were history buffs. But about 30 percent were also into fantasy and science fiction, and their number greatly expanded when there were actually fantasy and science-fiction games to be had. The second big burst of growth in the hobby took place in the late 1970s as a result of the widespread introduction of fantasy and science-fiction games.

The three other big forces to emerge in wargaming in the late 1970s were the gaming conventions, the increasing flood of gaming periodicals, and the publication of "serial games."

## THE CONVENTIONS ✱✱

As the hobby grew, the increasing number of gamers attracted an increasing number of publishers. It had become deceptively easier to get into the publishing business. This was aided considerably by the emergence of national conventions in the mid-1970s. The conventions had a few other effects also. The first "national" convention was Origins, held in 1975. Due to the vigorous promotion by Avalon Hill and SPI, it attracted a nationwide audience. Although the "nationwide" aspect was more publicity than reality, about 10 percent of the attendees did come from beyond "driving range" (a couple of hundred miles). Origins subsequently became an annual event, held either in June or July. The site was changed because nobody wanted to be stuck with the enormous amount of work necessary to put on the convention each year. Avalon Hill did the first two, SPI did the third in New York, and so it keeps moving, like the proverbial pre-19th-century army, unable to stop in one place too long, lest it exhaust the local resources and die of starvation.

The conventions, like most things in wargaming, were not similar to those of any other industry. There were a number of major differences. First, there were the manufacturers selling their wares at retail. This usually is not done. There are conventions where dealers will sell goods at retail prices, but not the manufacturers themselves. Not only did the manufacturers sell their goods at Origins, but they also used Origins as a convenient date to release new games. This practice has acquired a life of its own, and while some case can be made for it in a business and marketing sense, releasing new games at Origins has become more a matter of one-upmanship than anything else.

In addition to the dealers' area (which is probably the single-best-attended and most popular area at any convention) there are the seminar panels, lectures, and demonstrations. Here all of the "professionals" in the hobby get together, talk to, talk at, listen to, lecture to, and generally commune with their customers, the gamers. I have found the most remunerative approach to these seminars is to make them as two-way as possible. You can learn a lot from your customers. Gamers are not dummies, and working on games for a living tends to produce blind spots in the "professional."

The third unique quality of the conventions was the people playing the games—not only in the normal tournaments for prizes and awards, but also just open gaming. On any flat surface you could see, somebody would be playing some kind of a game—new games, old games, sometimes not even wargames. After spending the previous year studying games, or just giving them a quick look, this was the one opportunity many wargamers have to actually play them. The conventions also tend to be, as most conventions are, one vast social experience. I'm a shy person by nature, but even I got energized by running into so many interesting people. I ran around without a name tag on, as I'd rather talk to gamers as one gamer to another. I'm not interested in whatever illusion the gamer might have developed about who Jim Dunnigan is.

The average gamer is going to come to Origins with (as a per capita average) over $100. He's going to spend it all on whatever he sees at the convention. This is an ideal opportunity for a new company to buy a booth (for a few hundred dollars) and maybe sell a few hundred copies of the game. You get noticed and, more important, you get firsthand information on how your game is received. Also, you come to see other game publishers, especially those your size. You get to talk shop.

There are also several large regional conventions held every year, as well as a dozen or so smaller local conventions. The regional conventions generally bring in 1,000 or more people. It must be remembered that even Origins is essentially a regional convention that draws a certain number of people from outside the immediate area it is held in. It gets some attention from the national news media and fairly full participation by most dealers.

But, aside from that, Origins depends upon the local population for its attendance. Most of the conventions are held in the summer and all of them are primarily dependent upon local organizers.

While the conventions were a direct result of gamer participation and activity, the "professionals" of the hobby expressed themselves in one other area: publishing game-related magazines. *Strategy and Tactics* followed in the footsteps of Avalon Hill's *The General* (first published in 1964). About the same time the first issue of *S&T* came out (1967), others came out as well, some of which still exist. In the mid-1970s, *Fire and Movement* was begun. In 1972 SPI began publishing *Moves* magazine and in 1980 started *Ares* (a science-fiction gaming magazine). SPI even made an abortive attempt to publish a newsletter (*DataBus*) on computer wargames in 1975, but, obviously, it was a little too early at that point.

Also coming out in the mid 1970s was *The Dragon* (from TSR, primarily about role-playing fantasy games) and a number of other, smaller journals. These journals ("zines") usually provided the most lively and outrageous reading. They were truly personal but had very low circulations (rarely more than 1,000 subscribers). Even most of the other magazines rarely had more than 10,000 in circulation. The sole exception was *Strategy and Tactics,* which reached a peak circulation of nearly 37,000 copies per issue in 1980. The actual readership was more than 100,000, giving it the widest reach of any gaming publication. Most of the early magazines are still around, and provide the center for a very thinly distributed readership.

### The "Serial Games"

*Dungeons and Dragons* was more than an innovative and popular progenitor of a new game genre. It also owed much of its financial success to the "serial game" concept. Think of cameras and film. You buy a camera with a roll of film, and when that roll of film is gone, you have an urge to buy another roll of film, and another and another. Film sales are quite a bit higher than camera sales. *Dungeons and Dragons* operated the same way, with the basic game generating additional play value if you bought what eventually became hundreds of additional books, scenarios, figurines, dice, and so on. Avalon Hill was one of the few to do the same thing with a wargame. Their *Squad Leader* game, published in 1977, had additional sets of scenarios and playing pieces published every year or so up to the early 1990s. Also in 1977, Game Designers Workshop published its science-fiction role-playing game *Traveller,* which, like other role-playing games, generated enormous demand for additional materials. By the end of the 1980s, most other wargame publishers had done the same. The serial-game concept caused the game market to fragment into groups of gamers who tended to specialize in a particular game for the long term. Wargamers saw more of their members depart the fold because of this.

## INTO THE 1980s ✷ ✷

In 1979, what was a gamer to make of all this, and what were the gamers doing? Tastes were changing; so were the publishers—but not always in the same direction. The class of grognards was growing and changing and becoming more influential, while the new, larger generation of gamers were cutting their teeth on fantasy and science-fiction games, not the simple wargames and tactical armor games of the previous generation.

Gamers had changed from the late 1950s to the late 1970s, primarily because the games that they had been exposed to changed. In the early days, there were very few games. If you had any desire at all to be a wargamer, you had to like what was available. As more games were published, more people got interested. There was, so to speak, something for everyone. Beginning in the late 1970s, with the emergence of fantasy role-playing games and many more fantasy and science-fiction games in general, people who were not history buffs but had the mental capabilities to handle a game also became hobbyists. This has helped gaming in general, since many of the people who were not all that interested in history became interested through their exposure to fantasy and science-fiction games. This may, in a perverse way, demonstrate the most successful method of teaching history every discovered.

Gaming had also got out of the earliest stages of development, in which a handful of visionary (or sometimes just a little demented) people, perhaps a little too far ahead of their time, produced most of the games. Many players eventually lost their awe of the game-designing and publishing processes, decided to do it themselves, and, as a result, a far greater number of games were produced.

By 1980 the beginning was over. The hobby was not "new" anymore, but the most fascinating things were still to come. What made the developments of the 1980s so interesting was that most of them would happen on microcomputers. But while the next stage of wargame history had its share of innovation, it didn't have the kind of innovative excitement that existed at the beginning, even though the future would produce many far better games. There's nothing like being there at the beginning, with the opportunity to make a big difference. But for every beginning, there's an end, or at least a second act. The second act in this case wasn't nearly as much fun as the first one.

## THE DARK AGES AND RENAISSANCE

In the early 1980s, wargaming went through a period of tumult and subsequent renaissance. I saw it coming, but no one paid much attention. When I wrote the first edition of this book in 1979, I knew that changes were in the wind. Big changes. I knew this because SPI's customers had told me. One of my favorite innovations was to include an extensive questionnaire with each issue of *S&T*. This ran to 50 or more questions and was used to keep tabs on the demographics of our market as well as to test new product ideas and to measure the acceptance of existing games. We ran twelve of these voluntary-response surveys a year, plus one random survey for validation. I would spend one weekend a month running data through our minicomputer, performing all manner of arcane statistical analysis on the information received from our customers. A side effect of this system (called "feedback") was to make the gamers feel they were directly part of the decision-making process. They were, and what they were telling me in 1979 was disquieting.

Existing gamers were telling anyone who would listen that they were getting into microcomputers, and nonwargames. Although only 5 percent of wargamers had PCs in early 1980 (including myself, since 1978), the number was rapidly increasing. Given that PC ownership had nowhere to go but straight up, these PC owners were going to have less time, and money, for paper wargames. The second, and more ominous, trend was the increasing popularity of fantasy and science-fiction games at the expense of historical games. Interest was most keen for these games, especially role-playing games (*Dungeons and Dragons*) among the youngest gamers. Many new gamers were avoiding historical games altogether. In other words, the rapidly growing wargame market of the previous ten years had developed a bad case of clay feet. The older and more affluent gamers were wandering off toward PCs, and the younger ones were turning into dwarfs and trolls.

Just to make matters more interesting at this point, my partner, Redmond Simonsen, expressed a keen interest in selling out his share in the company. This was not the best time to sell, as SPI's recent major expansion into retail distribution had vastly increased the managerial workload and cash-flow requirements that produced an ugly balance sheet. However, it was a suggestion I could hardly ignore. So I scrounged up some buyers. Their offers all had two things in common: not enough money to satisfy the stockholders and a demand that I sign a long-term employment contract. A low selling price didn't bother me as much as five years of involuntary servitude. At that point (about when the first edition of this book was published), it dawned on me that it was time for another dramatic change in direction. I started SPI on a hunch, and by doing what no one expected created an opportunity for more of the same. Thus, I decided to leave SPI, giving my partners complete authority to sell the operation for whatever they could get. I could always make more money, and having left SPI, I couldn't be pinned down by a long-term employment contract. I couldn't make more time, but I could avoid losing it.

I warned all and sundry about the vast changes developing in the wargame market, but, as subsequent events demonstrated, many people in the wargaming business weren't listening, or they weren't listening hard enough. I then went off to do what I intended to do before getting caught up in running SPI for 11 years: write books. My first book, the first edition of this book, stayed in print for over 10 years. Not bad for a quickie on a "contemporary fad" (wargames). Before leaving SPI, I cut a deal with my publisher to do another book, and in early 1982 *How to Make War* was published and became a best-seller. This edition of *The Complete Wargames Handbook* is my ninth book, with several more in the works. It was a long detour through wargaming, but I finally got to where I was headed back in the late 1960s. During those last few years at SPI, I also discovered that it was more profitable modeling money than warfare, another new direction that paid off during the 1980s.

Meanwhile, the new leadership at SPI sought to exploit what they thought was the growing wargame market. Trouble was, the market was no longer there. It was fading fast. Instead of hunkering down to transform itself, SPI spent big in an attempt to capture a large market that was shrinking. A venture-capital firm had entered the picture during 1981, and was not happy with the way things were developing. In January 1982, the venture-capital outfit (which now had a large measure of control over SPI) asked me to take over running SPI once more. That was a tempting prospect, for about 15 seconds. I had other commitments at that point, and I knew that trying to turn SPI around in the then-current marketplace would be a daunting job. I turned the offer down, and by mid-1982, SPI's assets were acquired (in a rather complex deal) by TSR (the *Dungeons and Dragons* people). They got a bargain price, but didn't know quite what to do with it. In 1983, I got an offer from TSR to move out to Wisconsin and run its new wargames operation (what was left of SPI). Another offer that was easy to decline. I knew that it was only a matter of time before computers were owned by a sufficient number of people to enable wargaming to make a comeback. It had not happened by 1982, or 1983. It took most of the 1980s for wargaming to transform itself from a paper to an electronic medium.

The shakeout in the wargames business during the early 1980s was not a pretty picture. Consider the circulation of *S&T,* a good barometer of interest in manual wargames. When I left the wargame business at the end of 1980, circulation stood at 36,000. It was downhill from there.

| *Date* | *S&T* Circulation |
|---|---|
| November 1980 | 36,000 |
| November 1981 | 30,000 |
| November 1982 | 26,000 |
| November 1983 | 17,000 |
| November 1984 | 15,000 |
| November 1985 | 13,000 |
| November 1986 | 12,000 |
| November 1987 | 11,000 |
| November 1988 | 11,000 |
| November 1989 | 9,000 |
| November 1990 | 11,000 |

The slight pickup between late 1989 and late 1990 was partially my doing, as I returned to edit the magazine for a year or two (18 months, as it turned out). The reasons for editing the magazine once more were not too complicated. I got tired of gamers complaining to me about the constant decline in the quality of the magazine. When an opportunity to take over the editorship once more presented itself, I said, in effect, "Gimme that damn thing, now *this* is how you do it." The quality of the magazine (and to a lesser extent the games) had, indeed, declined after I left, and particularly after TSR took over in 1982. When I came back on board, I turned things around, at least to the extent that readers noticed a difference from issue to issue. But the major reason for the decline was wargamers becoming more interested in computer wargames and other types of simulations (fantasy and science fiction). All the talent was where the money was, fantasy, science fiction, and computer games. TSR had taken over publication of *S&T* in 1982, but was unable to make a go of it. In late 1986, TSR sold the magazine to 3W (a California wargame publisher). In turn, 3W sold it to Decision Games in early 1991. *S&T* is still being published by Decision Games, and 1992 marks the 25th year of the magazine's existence. The old rag is a survivor.

The relentless decline in wargaming activity during the 1980s was the result of a long list of factors. Including the ones we've already mentioned, they were:

- Older and more affluent gamers diverting their spare time into PCs and game software.
- Many gamers, particularly the younger ones (the "customers of the future") forsaking historical wargames for fantasy (RPG) and science-fiction games.
- The closing of over a third of the specialized game stores. Poor economic conditions in the early 1980s were largely responsible for most of this. Slowness in getting into software sales was another. The demise of SPI, which published an average of 27 new games a year, also hurt. In 1980, SPI's catalog listed 196 games in print. Many of these specialized game stores did a good business in selling these backlist titles.
- Changing lifestyles of gamers. Without the constant influx of younger gamers, the buying habits of the older gamers became crucial. But many of these men who took up wargaming in the 1960s were now raising families and getting into busy careers. Too many of the new wargames were complex (and expensive), and the main body of existing gamers now had less time and patience for the new games available.
- The precipitous decline of *S&T*'s circulation which denied the hobby a strong "center." The magazine played the role of media in providing a central voice for the wargamers who were otherwise spread thinly among the general population.
- The loss of analytic history in the shuffle. One thing I noticed when I began to edit *S&T* once more in 1989 was how much gamers had missed what I called "analytic" history. This was a method of presenting historical information developed when I first began editing *S&T* in 1969. Essentially, analytic history differed from the more common narrative history in that it, like the games, took a more numbers-oriented and "systems" approach. Analytic history attempted to link the various parts of an event together via the numerical data involved. I had noted that the editors who succeeded me after 1980 had drifted more and more into narrative history and, apparently, hadn't even noticed how much the analytic approach was missed. I found out rather quickly how enthusiastic gamers were to have their analytic history back in *S&T*. Unfortunately, the editorial skills needed to put together a piece using analytic history techniques appears to be rare. Over the years, I have had many individuals and organizations ask me to explain how to, in effect, "do analytic history" (including a memorable official visit from the CIA for that purpose). While it appears obvious to me, the knack apparently escapes most others who want to do

it. That's disappointing, so in an attempt to redress this problem, I will try once more, further on in this chapter, to describe how to do it.

As the *S&T* circulation numbers demonstrate, wargaming did not disappear; it merely contracted. There were still several major publishers. The major one was now Avalon Hill, which had hired many of SPI's product-development staff and set them up as a subsidiary game publisher (Victory Games) in New York City. Victory Games continued the SPI style and tradition of games, although on a much-reduced publication schedule. In 1983 Victory Games continued to demonstrate the old SPI ingenuity with a series of easy-to-play solitaire games. *Ambush* was the first of the series, and it used a paragraph book format to generate an intelligent and challenging opponent. This was a concept already present in fantasy gaming. Other publishers issued similarly easy-to-play solitaire games, and this was one of the reasons wargaming was able to survive at all. Unfortunately, Victory Games had a difficult time recruiting and retaining skilled staff. This was always a problem with the wargames industry, as it was basically a low-profit business that required very capable people (who could make a lot more money doing something else) to keep going. The remnants of the New York staff of Victory Games disbanded in 1989, although games under the Victory Games label continued to be published from Avalon Hill's Baltimore facilities.

GDW continued to operate in the Midwest, although it published a larger number of nonhistorical games in order to do so. There were several other companies of note that did not survive the early 1980s. OSG folded in 1982 and Yaquinto Games in 1983. Several other, smaller operations folded, but that was normal.

As wargaming companies finally took note of the substantial market changes going on, there emerged a growing movement to recast the image of the wargame industry, including the related offspring the role-playing games. This resulted in the concept of an "Adventure Gaming" industry. The idea actually began in the late 1970s when it became obvious that the role-playing games, although an offshoot of wargaming, were not only becoming a much larger business, but also quite a different one. In the early 1980s, this concept found expression in several mass-market magazines devoted to "Adventure Gaming" (*Adventure Gaming Magazine* and *Gameplay Magazine,* and in 1985 *Game News*). All soon folded, even though they were well produced. They tried to cover historical, fantasy, and science-fiction games in one magazine, and not enough people were interested. Specialized magazines were the way to go. TSR was quite successful with its *Dragon* magazine, getting circulation up to over 100,000.

Although no one was ever able to successfully compete with *S&T* (as a magazine with a game in it), there was a new magazine of that ilk begun in Britain in the late 1970s, *The Wargamer* (publisher by World Wide Wargamers, or 3W for short). Rather crude by *S&T* standards, it was serviceable and enjoyed some success in Britain, and in the early 1980s the operation was moved to California. In the wake of SPI's demise, *The Wargamer* improved its quality and frequency of publication. While *S&T*, and previous competitors, had been bimonthly, *The Wargamer* reached monthly publication by 1986. But this was pushing it. It was difficult to get a new game out every month. A novel solution to this killer schedule was achieved when TSR began shopping *S&T* around in 1986. There were eventually several bidders, and 3W acquired *S&T* in late 1986, merged *The Wargamer* into it, and began publishing *S&T* eight times a year. Even with a circulation of 12,000 in 1986, *S&T* was profitable. Profits were enhanced, ironically, through the use of PCs. Microcomputer technology was now capable of handling layout and production of the magazine. The low overhead of 3W's rural California location also helped.

Although the circulation of *S&T* had declined 75 percent from the glory days of the late 1970s, the number of wargamers had shrunk a bit less. Those games that were distributed through stores sold about half as many units per titles as in the past. Publishers adjusted for this by raising their prices. Because most of the remaining wargamers were well-educated, older guys with high incomes, the price increases did not have much effect. Sure, these men still didn't have much time. In fact, surveys in the late 1980s indicated that 40 percent of them simply "studied" the games they bought, only a third actually played, while the remainder put them on the shelf for future reference. This is remarkably similar to what happens to a lot of books bought. The publishers really don't care, unless the customers stop buying.

Several novel methods were adopted by wargame publishers in an effort to increase sales. Many of these ideas worked, but only to a limited extent. Large and complex games could be priced quite high, with hefty profit margins built in. But if one of these didn't achieve the usual (low) sales, there was a significant loss incurred. GDW and Avalon Hill tried publishing very simple games. But these were intrinsically low-priced games, with little in the way of generous profit margins, and you had to move a ton of them to make any financial progress. The only games of this sort that sold respectable numbers were the movie tie-ins. Avalon Hill did well with a *Platoon* game (after the movie of the same name). Victory Games did a line of James Bond role-playing products. Ironically, Victory Games's greatest commercial success was a *Dr. Ruth Sex Game*. But this was not wargaming, and the publishers were aware that they were getting away from their reason for being by publishing games that were much removed from historical simulation.

TSR, which had acquired the rights to SPI's backlist, made steady, if modest, profits by republishing some of the more popular SPI titles. This is not a big money-maker, but it has made wargamers happy by keeping some of the classics in print. More profitable were the tie-ins, especially games like *The Hunt for Red October*. TSR also published some original games under its SPI trademark, but this has been a minor part of its publishing activity.

Inexorably, computer wargames became more prominent, with reviews and ads for computer games becoming increasingly prominent from 1984 on. *S&T* regularly featured full-page color ads for computer wargames. An encouraging development was the increasing use of historical material in the games' instruction manuals. By the late 1980s, some games had separate "data manuals" running to 100 pages or more. This appealed immensely to the historical gamer, who was essentially an information junkie. By the early 1990s, a lot of this information began to appear within the games themselves. This was possible because most games now required a hard disk, which meant that thousands of words of text (plus illustrations) could be stored on the hard disk without interfering with the play of the game.

History had also caught up with historical games. After SPI published the first game of NATO/Warsaw Pact combat in 1972 (*Red Star/White Star*), this new genre became the best-selling of historical periods, even though it was "history" that hadn't happened yet. But the most salable games always featured Europe and a clash between NATO and Russian forces. When peace broke out in Europe with the collapse of Communism in 1989, interest in games of this type also collapsed. At the same time, interest increased for games on older periods, particularly premedieval history. Wargames went back to their roots.

## ANALYTIC HISTORY, AND WHAT IS A SIMULATION ANYWAY? ✷✷

Two things that defined the wargaming scene in the early days, aside from the flood of games, were the use of the term "simulation" rather than "wargame," and the unique nature of the history articles in *Strategy and Tactics* magazine. There were interconnected reasons for both of these developments. The primary reason is that wargamers are not so much gamers as they are very curious history buffs. While many military-history enthusiasts are content to read a book on battles and campaigns, watch a good film on war, or wander through battlefields, a wargamer wants to measure and analyze things. When I started SPI, the name, Simulations

Publications, was chosen as an accurate description of what we were doing, and what we felt wargamers wanted. What we were creating weren't wargames, they were simulations. We were stuck with the traditional term "wargame," but every one who created or used our "games" knew better. We called them wargames, and let others call them wargames, an incorrect term for an activity that was incomprehensible enough as it was. It wasn't worth the effort to expunge inaccurate terms, although we tried. Wargamers used the terms "wargame" and "simulation" interchangeably. But the labels issue was small change compared to the emergence of analytic history.

Analytic history is what a wargame was before it became a game. A wargame is, after all, a historical account of an event in simulation form. The subject must be researched and the data organized so that it can be presented in a simulation format. That's analytic history. But wait, what the hell is the difference between a wargame and a war simulation? Often there's not a lot of difference at all. The main purpose of a simulation is to present the situation so that you can manipulate the key elements. This allows you to better understand how all these elements interacted, and this is where the game element comes in, the opportunity to play around with alternative strategies and tactics. This explains why so many wargamers don't game at all, but simply study and manipulate the game by themselves. Yes, two people can use a simulation as a game—many do—and the game element is not ignored when putting these things together. The game element is there whether you want it or not. It's the nature of the beast. There are usually two sides in a military conflict, both have numerous elements of the situation they can manipulate, thus you have a game situation. Some wargamers enjoy the game element more than the simulation and history aspects. But most are basically into the history; otherwise they would play the more numerous nonhistorical games. And that is what many early wargamers did when the fantasy and science-fiction games came along. These games were more game than simulation, although a simulation element was present.

It was easy to miss the concept of analytic history in all this. Few gamers saw the research material that went into the game. When I developed the concept of having an analytical article accompany each game (first with games in *S&T* magazine, later in the boxed games), most gamers simply noted that the historical material was well organized and complemented the game well. The analytic history was taken for granted until it wasn't there anymore. Even then, gamers weren't sure what they were missing, except that whatever it was, it didn't seem to be there anymore.

*This* is what was there: in other words, a concise description of what analytic history is and how to do it.

First, you have to see the historical event as made up of several discrete parts. These parts are:

**The Components—** Just like in a game, you have essential parts of the history. These, like a game, include the map, Order of Battle, combat results (what caused losses and to what extent), terrain effects, and the "sequence of actions" (the Sequence of Play in the game). These are treated as separate, self-contained, miniarticles.

**The Time Line—** That is, a description of what happened when and who did what to whom and with what means. This is the familiar narrative or descriptive history. This usually serves as the main body of the article, with the other components hanging off it as sidebars (or "modules," as I called them, as I didn't know what a sidebar was at the time).

**Presentation—** Ideally, you put as much graphic design talent into an analytical-history article as you put into a game. Analytic history is as much an organized, graphic device as a wargame is.

**Selectivity and Conciseness—** Just as the games provide value by compressing a lot of information into a small space, so must analytic history. Just as a well-designed game is selective about which elements of the historical events are covered, so must analytic history be. This selection process is probably the one thing most would-be analytic history authors and editors have trouble with. I suppose the best way to hone your selectivity skills is to design a number of games. Turning a historical situation into a manageable game forces you to be selective. Many attempts at game design are unmanageable because of a lack of selectivity, which is not good. Selecting which items to cover in a 3,000–6,000-word article, and which to give the graphic treatment, is the key. The problem is frequently present in games, where the designer will include too much detail on some aspect of the situation that really doesn't warrant it—having overly complex rules for artillery or air power are common errors. Selectivity is a rare skill. Some have it, most don't. As with learning game design, you have to practice. There are no shortcuts.

Analytic history has been done before, although not deliberately. Treating accounts of historical events in a systematic and highly organized fashion is not unique, but the relentlessly organized approach of analytic history is. If you do it right, the reader gets a large dollop of knowledge for a small investment of time. What's more, the reader doesn't even know he's been hit with a carefully crafted presentation the creators know as analytic history.

A historical simulation attempts to duplicate a past event, including duplicating the key elements of that past event that the original participants had to deal with. Most wargames, including chess, do this. What makes a simulation such a powerful form of communication is that it is, like most events, nonlinear. A book or film is linear. The author leads you from point to point, with no deviation allowed. Simulations, games in general, and analytic history, are nonlinear. That is, you can wander all over the place and still be somewhere. Flip through a book, and you pick up pieces out of context. Make different moves in a game, and you have a context, because the game allows, even encourages, deviating from the historical events. Linear media can be a drag at times; nonlinear media keep you on your toes. Analytic history is written with nonlinear use in mind. You can wander around a piece of analytic history and still get a lot of useful information.

Simulations, and to a lesser extent analytic history, require a fair bit of effort on the part of the user. Film is more popular than print because film requires less effort to get something out of it. Similarly, you don't have to be terribly observant in a bookstore to see which history books sell best. History books with a lot of pictures are more popular than those with a lot of words. Print is more popular than simulations for the same reason.

The dozen or so marketing surveys I analyzed each year at SPI also demonstrated the greater popularity of the simple over the complex. But toward the end of the 1970s, and into the 1980s, a contrary phenomenon expressed itself. It seemed that those people who stayed with wargaming tended to prefer more complex games. Thus, if a publisher wanted to hang on to the existing market, more complex games had to be produced. Yet it was also obvious that new gamers needed simpler games. Going into the 1980s, the market for simpler games was shrinking while the market for complex games was holding its own. Actually, the complex game market was also declining, but it was declining a lot less. Publishers continued to put out some simple games, and even the most avid and experienced gamer liked to deal with a less daunting game from time to time. The changing lifestyles of gamers actually led to their using the more complex games like books: as sources of information. There was little time to actually set up and play the more complex games. So a well-done simple game was still welcome as a game that took little enough time and effort that one could actually play it.

Now in my experience, there are two ways to make the most of a simple game. The best and, naturally, the most difficult approach is to sweat blood in order to develop simple but elegant rules to accurately reflect what was going on in the situation. The other approach is to concentrate relentlessly on playability, to cut corners with historicity as much as you can get away with it and hope the gamers won't notice. Well, some of them will notice, and some of them won't. In either case, it helps if you have good graphics and good play-testing. All of this may be moot, however, as computer wargames continue to replace paper wargames of *any* degree of complexity.

## WARGAMERS OVER THERE ✷✷

While modern commercial wargaming got its start in America, it quickly spread to other countries. The reason for the migration of wargaming was quite simple: Most wargamers are well-educated people with an interest in history and technical matters. Every nation has such a "technical/professional" class, and these were the ones who picked up on wargaming. The English-speaking nations were, obviously, reached first. Britain already had thousands of miniatures gamers and an even larger number of avid military historians. Outside the United States, history is a much more popular indoor sport. Except in Germany, military history has no stigma attached to it and, in general, there were many potential wargamers available overseas. Miniatures wargaming, the earliest form, began in Europe but never became popular because of its complexity and expense. Those who did wargames over there were quite dedicated and often went beyond wargaming in their military curiosity. One of the best examples is Fred Jane, a turn-of-the-century British naval miniatures wargamer. He eventually began publishing his wargames research as *Jane's Fighting Ships*. That volume, published annually for many decades, has led to similar annual volumes on everything from aircraft to shipping containers. Jane's publishing company is currently the largest publisher of military information in the world. And all because Fred Jane wanted to develop more accurate rules for naval wargames.

The big impediments to the spread of wargaming overseas are money and language. The British wargamers have little problem with the language, although American military terms are somewhat different from British ones, and the wargame terminology has to be learned in any event. Money, however, was and is a big problem. Among the English-speaking nations, Britain is actually one of the poorest in terms of disposable income. And then there are the import duties. Making adjustments for all these factors, a $20 wargame bought in America is more like a $50 purchase for a Briton. In the early 1970s, I set up an independent agent for SPI games in Britain, where eventually we were able to manufacture many games. This brought down the cost considerably. This British outpost disappeared in the mid-1980s, but it had already served the purpose of building up the numbers of wargamers in Britain and making it easier for British wargaming companies to get started.

Australia, with a smaller population than Britain and with a much smaller SPI agency arrangement, managed to develop an even stronger local wargames industry. Maybe it was the warmer weather—who knows? It may have had something to do with the weaker tradition of miniatures wargaming and a more colonial attitude of just jumping into a new concept and pushing it as far as possible.

Canada, to the everlasting chagrin of many Canadians, was just an extension of the U.S. wargames market, although a number of small wargames publishers developed up north. SPI did have the dubious distinction of publishing the only U.S.-manufactured wargame to sell more copies in Canada than the United States. The game was on the Canadian separatist movement of the 1970s. It was called *Canadian Civil War* and had the additional questionable honor of having the first sample copies seized at the border by Canadians customs officials as "seditious material." Great publicity it was, too.

The non-English-speaking nations had to overcome the additional problem of language. It wasn't just a matter of translating the game rules from English to the local lingo; we had to come up with workable new words in the foreign language for the unique technobabble found only in wargames. This was a major problem for many years, although many nations ended up doing what they have done with so many other English technical terms: They just took the English terms and used them as is.

Most industrialized nations have a high percentage of English speakers among their professional/technical classes (often over 80 percent, as in the Netherlands). In these nations, many wargamers simply bought the American version and played it. Eventually, enterprising local gamers began to develop games themselves and publish them in their own language. This was in addition to licensed translations of best-selling American wargames. Some nations, like Japan, have translated locally designed and published wargames into English and sold them in the United States.

In most European countries, and Japan, wargame publishing began as a sideline for established toy and print publishing companies. This meant that the products were, well, a bit more subdued than in America, where there was a tendency toward more flash and less substance. Nevertheless, there were some outstanding wargames put out.

By 1983 there were five wargame publishers in Japan. But through the rest of the 1980s, fantasy and science-fiction games began to edge out the wargames. Most of the wargame-publishing operations cut back or shut down. The same thing happened in Europe. One interesting development in France was the emergence of one of the few gaming magazines that successfully combined historical wargames, fantasy, science-fiction, and computer games. The magazine, *Casus Belli,* is owned by a major publisher, has a circulation of over 50,000, and is quite profitable.

There is even wargaming in many Third World nations. The major restriction there is money. Wargames are relatively expensive. Much of the interest in these nations comes from the professional military. U.S. professional wargames are quite expensive, and few nations of any size outside the United States can afford them. In fact, it was customary for the U.S. military to send friendly foreign staff officers to me to obtain "wargaming they could afford." This brought some interesting visitors to my doorstep, including a senior general of the Egyptian General Staff. In one case, I was visited by some U.S. officers, sent by the U.S. general commanding U.N. forces (U.S. and South Korean) in Korea, for some advice on how to get the South Koreans into wargaming.

There were some even more curious incidents. Once, while at SPI in the late 1970s, I was visited by FBI agents. They wanted to look at our mailing list, as they believed someone was buying wargames from us and illegally exporting them to mainland China. They never told me if they caught the, er, smuggler. The Nationalist Chinese (Taiwan) were more up-front. They would buy games (apparently legally) and several times took me out to dinner to talk shop. Well, at least they were on our side.

The Russians were also great fans of U.S. wargames. They were particularly interested in World War II games, and senior staff at the Russian U.N. delegation were quite eager to get their hands on SPI's *War in the East* game. I eventually did find out who they had play the German side. In 1989 I was invited to dinner by a Russian Navy captain who had worked on wargames in the Russian staff academy. He pointed out something I already knew, that Russian wargaming was either miniatures or simply a series of formulas. At least that was the official form of wargaming. He allowed as how paper wargames were eagerly played whenever Russians could get their hands on them (money again). Oh yes, he noted that the junior officers had to play the bad guys, and the junior officers were expected to show proper respect for their superiors (who played the Russians). Some things never change.

As PCs began to proliferate overseas, computer wargames became popular. If anything, computer wargames became more popular in Japan than in the United States. Going into the 1990s, the computer wargame is the growth segment of wargaming overseas. Paper wargames still survive outside the United States, but in a much diminished form.

## HOW MANY WARGAMERS ARE THERE? ✱✱

There aren't enough of them, that's for damn sure. Sales patterns indicate that at its peak in the late 1970s, there were only a few hundred thousand historical wargamers in the nation. There were about as many throughout the rest of the world. As of the early 1990s, sales patterns indicate that there are probably only about 100,000 paper gamers still active. Computer wargames are another story, with several hundred thousand regular devotees of the genre and the number steadily growing. By the end of the decade, there will probably be over half a million. This growth is made possible by the increasing ease of use of computer wargames. Granted, most of these new computer wargamers are playing simulator-type wargames, but they are just as eager for historical simulation and the historical accuracy that goes with it.

There are about 10,000 miniatures wargamers worldwide. These guys spend a lot of money, so their market clout is larger than their numbers would indicate.

Wargaming is largely a male avocation. There are women wargamers, but they comprise only about 1 percent of those active. The figure is somewhat higher with computer wargames, perhaps 2 or 3 percent. Aside from men being more curious about warfare, the only scientific research I could ever find on this issue was that one experiment showed that women prefer games with more (apparent) chance while men prefer games with more (apparent) skill. Well, I guess that explains why you find casinos with more women playing the slot machines and more men playing card games.

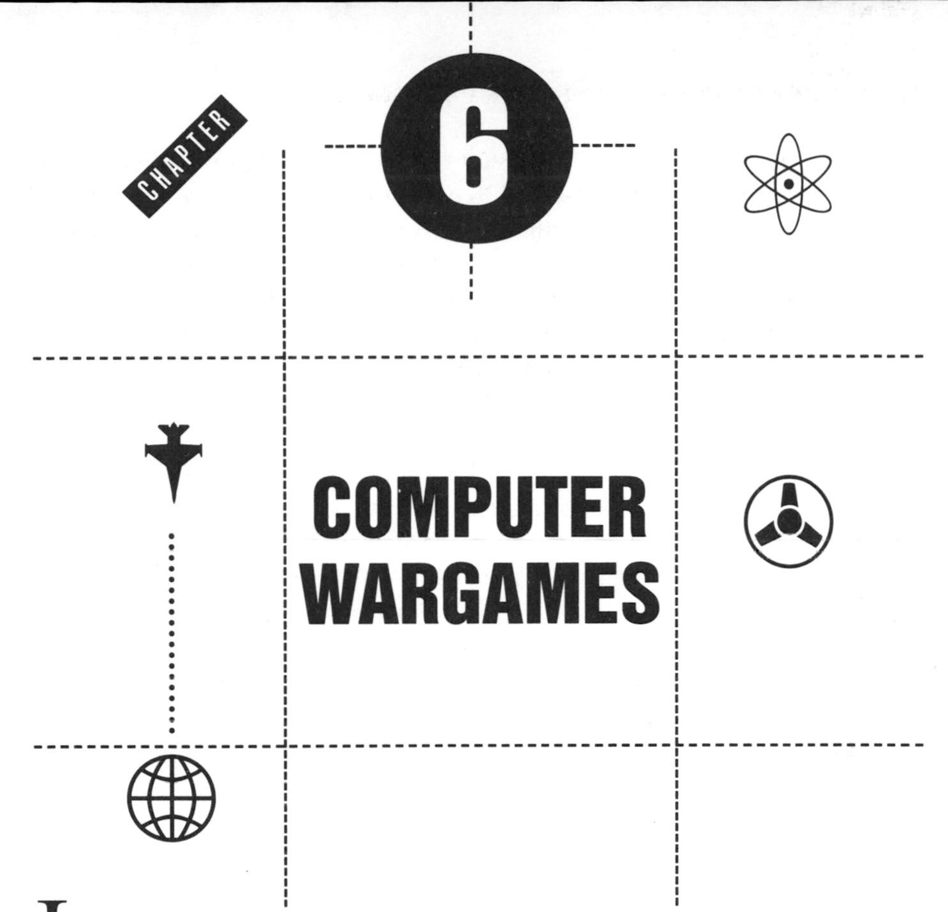


# COMPUTER WARGAMES

In the past 15 years, computers have increasingly displaced paper as the format of choice among wargamers. It has not been an easy transition. In the long run, computers are leading to yet another golden age of wargaming. But in the short run, computers have not had a very comfortable relationship with wargames.

In the early 1990s, about two thirds of all manual (paper) wargamers have personal computers. About two thirds of *this* group actually play computer wargames. Yet while annual sales of manual wargames amount to a few hundred thousand units (and dropping each year), computer wargames sell over half a million units a year (and are rapidly increasing). If you include the simulator-type games (aircraft, vehicle, and others), several million computer wargames are sold each year. Computer wargame sales keep climbing as more powerful computers make these simulators even more impressive, easy to use, and attractive to people who would previously not have wanted to hassle with the traditionally (and unavoidably) complex wargames.

As personal computers become more powerful, the games that can be written for them become equally more impressive. But wargames account for only about 10 percent of all computer games sold. About a third of the computer games sold are role-playing games (generally of the *Dungeons and Dragons* variety). Another 20 percent are action/arcade type games, despite the competition from the cheaper Nintendo-type game machines. And about 25 percent of games sold are of the simulator variety, usually putting the player in the cockpit of an airplane. The future of gaming obviously, is in computer wargames. You can get a better idea of where that future is heading by taking a look at its recent past.

Until the advent of microcomputers, any use of computers in, or with, wargames was largely hypothetical. But in 1977, when the first personal computers (PCs) appeared, computer wargames had a ready market. At that time, many gamers (more than 5 percent) worked with computers for a living, and twice as many again had access to a large computer in the course of their work. And even at that time, when commercial-computer wargames did not exist, a lot of computer time was spent on (rather primitive) wargaming on office mainframes and minicomputers. Many of these games were very interesting, and a lot of the work that was done on them was soon transferred down to microcomputers.

Through the 1980s, more and more wargamers got themselves PCs. By the early 1990s, over two thirds of those who played manual wargames also had access to a PC. And all of those people playing other computerized games had PCs, and an easy way to get interested in (computer) wargames.

## WHAT KIND OF COMPUTER, WHAT KIND OF WARGAME? ✷✷

What kind of computer attracted the first computer-wargame publishers? At first it was whatever was available, and these were the primitive machines that first appeared in 1977. Basically, these were machines you could play a game on, and were relatively cheap. Initially, most wargame software was written for the Radio Shack TRS80 Mod I. Radio Shack managed to get the jump on everybody else in that they got more PCs into more people's hands than any other single manufacturer. There were more than 200,000 of them in use by 1980. Nearly 5 percent of all wargamers had them. There were more TRS80s at that time than Apples (which were even better game machines) and most other microcomputers combined. But the situation soon changed.

The PCs of that era were little more than a keyboard, a TV screen-type device (in some cases, it actually was a TV), and a tape recorder that was used to load in programs. This whole outfit could be had from Radio Shack for $850 as of early 1980. For that price ($1,400 in 1992 dollars), you can now get a 486 with four megabytes of system memory, VGA graphics, and a 100-megabyte hard disk. The 486 is over 60 times faster than the TRS80, and can display much more realistic graphics.

The 1980 PCs had very poor graphics, especially the TRS80 (which was only black-and-white). One reason why the Apple II and Commodore 64 PCs eventually surpassed the TRS80 was because of still crude, but much more effective, color graphics. Despite the lack of striking graphics, there were some wargames on those early 16K and 48K machines. Note that the "K" refers to system memory in thousands of bytes (or characters) of information. Many current games require 2,000K for games, or 40 times as much as was available in 1980. The earlier machines required programmers to make every byte count, and they did. But these games were not all that attractive.

For most of the early games, the "game" generally consisted of asking you a series of questions. Even flight simulators operated in this fashion, and they sold. Then as now, most computer simulations were solitaire programs, even though some of the earliest ones tried the two players (on one PC) route. These solitaire games did not bother wargamers, as most gamers normally play their games solitaire for reasons discussed earlier. The solitaire aspect of computer games was in fact an asset.

The big change came in 1981, when IBM ("Big Blue" in the trade) introduced its PC. The IBM machine initially contained up to 256K and had larger-capacity floppy-disk drives. Although many early PC owners used the cassette drive to load software, most quickly converted, by the end of the 1970s, to floppy disks. But until the IBM PC came along, floppy disks generally held less than 100,000 characters of information. The first IBM floppies held 160,000 characters and soon increased to 360,000. By 1990, the average floppy disk held over a million characters. Today, mass storage, in the form of sealed "Winchester" drives, holds 100 million characters and up while costing less than the 1980 floppy drives.

By the mid 1980s, many manual (paper) wargames had been directly transferred to computers. This meant displaying a hex grid on the computer screen. While the screen was much smaller than the traditional game maps, this was got around by allowing the computer-game player to look at different portions of the entire map. In effect, the computer screen became a window on a larger map. While the small screen was one disadvantage, there was an even larger problem with these computer wargames. The player had to use the keyboard and/or mouse device to move the playing pieces around, which proved very tedious. On the positive side, the player did not have to learn the game rules, as most of the procedures were embedded in the computer program. The player did have to figure out how to use the program, but computer wargames became easier and easier to use as programmers learned more about what players were most comfortable with.

One thing wargamers were not comfortable with was the inability to know exactly what was going on inside the wargame. One major advantage of the manual games was that all the numbers and procedures were right there in front of you. Not only could the player see how the game did its stuff, but the player could, and often did, change things he did not agree with. This was not possible with computer games, although this is developing somewhat as we enter the mid-1990s.

Despite these drawbacks, computer wargames attracted more and more attention. In the late 1980s, a majority of the wargamers were playing computer wargames, and an increasing number of them were not playing manual wargames at all. Part of this trend had to do with the shrinking number of manual wargames caused by the introduction of role-playing games and computers and made worse by the remaining gamers getting older and having less time to play manual games. But as these gamers got older, they had more money and could afford to buy several generations of PCs and the games to go with them. The time factor for playing manual games, and the fact that they could not be easily left sitting somewhere to be finished later (animals and small children easily upset these games in progress), gave computers another edge. While there will probably always be some manual games, computerized wargames are definitely the format of the future.

Computer-wargame design has been driven by technology. This creates a much more vigorous publication schedule for computer wargames. Every few years there is a dramatically new generation of machines to run the games on, and topics are recycled in newer and more impressive programs to take advantage of the new technology. Through 1980 there was very little computing power available, and only the most primitive graphics. When IBM entered the market in 1981, it cut short Apple's increasing dominance of the market. The IBM machine was not that much of an improvement over the 1977-era Apple II, but the IBM machine had more growth potential and IBM's marketing clout enabled it to sell more PCs. By the mid-1980s, IBM machines had high-resolution graphics and a new generation of machines (the "AT"). About that time, game publishers found that gamers would pay for wargames that took advantage of new hardware. In the late 1980s, publishers found out that the core of their game-buying audience was increasingly buying new hardware specifically to play the latest game releases. By 1990 publishers were abandoning the old "lowest common denominator" approach to computer-wargame design and publishing games that couldn't even be played on over 10 million older 1981-era IBM PCs. The hard-core, enthusiastic wargamers were buying so many wargames that by the early 1990s, publishers increased the minimum hardware for many of their new games on an annual basis.

Today, many new games require a state-of-the-art machine (a "486/50" type) for best performance. The games produced under these conditions are spectacular. The genre that has benefited most is the simulator. This type of wargame puts the player in an aircraft cockpit, or tank turret, and allows him realistically to operate the equipment. Crude versions of simulators have been available from the earliest days of PC wargames. But as the PCs have become more powerful, the simulators have come into their own. There had been manual simulator wargames, but these were never very popular because of the enormous number of rules to be mastered in order for the simulator to be accurate and convincing.

In effect, software (the games) follows hardware (the machines to play them on). That is, computer-game features are limited by what the computers they run on are capable of. In 1984, the "eight-bit" PCs ruled. These were the first-generation PCs, first released between 1977 and 1981. In 1984, 19 percent of the games sold were for the Apple II (first shipped in 1977), and 51 percent were for the Commodore 64 (1981). The Commodore machine had more games sold for it because the Commodore PCs sold for less than half what an Apple II cost and thus had a lot more machines out there. In 1984 games for IBM machines accounted for only 8 percent of computer-game sales. The new Apple Macintosh (like IBM, a second-generation 16-bit machine) also had 8 percent of the game market. Four years later, in 1988, the situation was radically changed. At that point, 47 percent of the game software sold was for IBM machines. The reason was simple—there were a lot more IBM (and clone) machines around. While at the end of 1984 there were only three million IBM PCs in use (in the United States), by the end of 1986 there were over six million, and by the end of 1988, nearly ten million. Although the IBM machine was intended primarily for business use, an increasing number went into homes, and a lot of games were played on the office machines (which is why an increasing number of computer games had a "boss key" that, when tapped, put a dummy spreadsheet on the screen). Today, there are 25 million IBM-type PCs in the United States, with about half of them at home and nearly two thirds of all machines being used for game-playing at one time or another. After word processing, games are listed as the most frequent use of machines in the home (about 50 percent of those surveyed admit to playing computer games).

But it wasn't just numbers of machines that counted, but what they could do. Apple's new Macintosh computer was, in many ways, more capable than the IBM. But Apple made a mistake in not having a color display right away for the Mac. In 1985 the equally advanced Amiga and

Atari ST appeared (with color displays), but these were buried by the IBM avalanche. When IBM introduced high-resolution graphics in the mid-1980s, IBM PCs became the most powerful and widely available machines to put games on. As more IBM owners upgraded their older machines with the higher-resolution graphics, and more new machines were bought, the IBM market became huge. By 1990, publishers could afford to invest up to a million dollars developing a highly popular computer role-playing game and still make money on it.

## THE PERILS OF COMPUTER WARGAME PUBLISHING ✱✱

Spending hundreds of thousands of dollars developing each new game did lead to a number of computer wargame publishers going bankrupt. The largest manual wargames never cost more than a tenth of that. Computer games are more expensive because the people working on them get paid a lot more. These higher pay scales exist because your average game programmer could be making even more money working on nongame software. Games are fun, so many programmers take the pay cut. But even then, most experienced game programmers are making $50,000 a year and up. The people designing manual wargames make half that, or less.

Computer-game publishing was and is a competitive business, even if you can get away with paying the authors (programmers) minuscule advances on royalties and paying your programmers less than any other (nongame) employer of programmers. If the games don't sell in sufficient quantity, you lose money. This problem is compounded by two other killer items: returns and other types of computer games competing for the wargamers' attention (and dollars).

Returns are the policy whereby stores buy games from the publisher with the understanding that they can ship back what they can't sell. The bright side of this is that the store owners don't have to take on a large financial risk to carry a game; the downside is that publishers in these circumstances do. If too many unsold games are returned, the publisher takes a big fiscal hit. The returns are often a total loss, or can be resold for only a fraction of their manufacturing cost. Oddly enough, returns were much less common among manual wargame publishers. But everyone smelled more money in computer-game publishing, and "return privileges" were universally adopted as a competitive tool. Now nearly all computer wargame publishers are stuck with a practice that, for many of them, has proved fatal.

Publishing computer wargames, or any other computer game, isn't all that complicated. Let us examine the numbers. For a typical low-budget computer wargame in the mid-1980s, the publisher could get away with giving the designer/programmer $10,000 (or less) and then take delivery of the game, print a minimal manual and a flashy box cover, duplicate the disks, and ship the game. In those days, the game would fit on one disk, so the total cost of goods (disk, manual, and packaging) was two or three dollars. The game would list for $30, and discounts to distributors would leave the publisher with $12–$14 per game sold. Depending on how lucky the publisher felt (and how large the prepublication orders were), 5,000–10,000 units would be manufactured. Thus, for the $10,000 advance on royalties, $15,000 manufacturing cost, and maybe a few thousand dollars for ads, the game would be (as they say in the trade) "in the (distribution) channel." If all 5,000 were sold and paid for, the publisher would realize about $60,000. Royalty rates on such a deal would run about 15 percent. That would be $9,000 at this point, so the designer would still not get anything, as the advance was $10,000. So the gross profit would be $30,000 ($60,000 less $10,000 advance, $15,000 for packaging and shipping 5,000 units, and $5,000 for advertising). A typical small publisher in those days would need $250,000 a year to keep a few people on staff, pay rent and expenses, and so on. If six games were published a year, each game would have to cover over $40,000 in overhead. So selling 5,000 copies wouldn't make it. Even if the publisher could sell 10,000 copies ($120,000 in sales less $18,000 in royalties, $30,000 in packaging, $7,000 in advertising), the gross profit would be only $65,000. Less the $45,000 overhead and there's only $20,000. That had to cover games that bombed completely, games that required more work in-house or took a lot longer to finish (very common), and distributors that went bankrupt and left companies unpaid (which happens quite a lot in this business).

What made it so difficult for wargames to get decent sales was the competition from, well, more entertaining software. The first computer games were essentially eye-hand-coordination games. These are better known as "arcade games" (they appeared in arcades in 1975, before PCs were common). The wargame simulators are direct descendants of the arcade games. While tactical and strategic skill are required in combat simulators, nimbleness with the joystick is paramount. And then there were all those science-fiction and fantasy games (including the sexually exciting ones, like *Leisure Suit Larry* et al.). Most publishers realized that they could make more money for their efforts by working on simulators or fantasy titles. This left wargame publishing to the smaller outfits who had even less clout and influence in getting their games into a lot of stores.

Fortunately, there were sufficient numbers of wargamers available to encourage the publishers to keep publishing. Going into the late 1980s and early 1990s, however, the publishing landscape began to change considerably. Now publishers didn't just *want* a hit best-seller, they *needed* one to stay alive. But this story is still developing.

## GENEALOGY OF COMPUTER WARGAME TECHNOLOGY ✱✱

One way to get a better handle on where computer wargames came from, and where they are headed, is to take a look at the development of computer wargames in terms of their computer technology. Chapter 5, on the history of wargames, covers this subject in a more general way, but here is where you find the juicy details.

Not surprisingly, the first computer wargames were developed unofficially by students at universities specializing in computer research. The earliest-known game of this type was *Space War*, in 1961 by Steve "Slug" Russell at M.I.T. About the same time, the army was developing *ATLAS*, the first of the theater-level wargames on the possible war in Europe between NATO and the Warsaw Pact. The big difference between *ATLAS* and *Space War* was that *Space War* was more of a game, with the player being constantly involved. *ATLAS* was more of a model; you set it up and it played out the situation without human intervention. ATLAS was created not by students but by professionals. *Space War* was the ancestor of all PC-based wargames, while *ATLAS* was something of a dead end.

In the late 1960s came the first of the "Adventure" series of wargames. These were a combination of role-playing games (which hadn't been invented yet) and puzzles that were only tangentially wargames because they involved combat and the kind of strategy successful generals are supposed to be capable of. These were text-only games, whereby the program presented the player with a "world" to move around in and a lot of prizes to grab and dangers to avoid. Also in the late 1960s, the first *Star Trek* game appeared, although this was basically a spiffed-up version of the earlier *Space War*. Also found on many mainframes and minicomputers was an artillery game that, like *Space War* and *Star Trek*, used crude graphics and a lot of math to depict combat situations.

The first computer wargame available to the public was Nolan Bushnell's 1971 *Computer Space* (a version of *Space War*). Mr. Bushnell later unleashed *Pong* and *Asteroids* on the public. These were extremely primitive games run by about as much computing power as you'll find in a 1992 digital watch. Many younger gamers today have never even seen these early games, other than the glitzy variations that appeared in early Nintendo games.

In the early 1970s, young wargamers like Chris Crawford attempted to convert commercial manual wargames to computer versions. But the hardware, and to a certain extent the development software, just wasn't there yet. I spoke with many gamers in the early 1970s who were trying to create computerized wargames. I was well aware of the basic problem, which was the cost of the machines to play the games on. The introduction of PCs in 1977 resulted in an astounding decline in the cost of computing power. In 1973 I leased a minicomputer that had 16K of memory and used punched cards to get information into it. This device cost $136,000 (1992 dollars). Five years later, you got four times as much memory and keyboard input for under $5,000.

In those early days, much of the action was at places where the computer resources were, essentially, "free" to developers and users alike. One such location for wargamers who were programmers (including Chris Crawford), was the PLATO project, a multi-user educational system then under development at the University of Illinois (at Urbana) and heavily supported by supercomputer manufacturer CDC. Begun in the late 1960s, by the mid-1970s there were up to a thousand users on the system at once. The system was quite advanced for its time, with graphics equal to the early Macintosh computers 10 years in the future. I was invited out to the university in 1969 to give a lecture and wasn't all that surprised that many wargamers were working on the system, or that wargames would appear on PLATO eventually. During this period, a number of innovative games were developed for the PLATO system, including a 3-D *Star Trek* game and more elaborate versions of *Adventure*. By the late 1970s, PLATO had multi-player games, something that did not see commercial availability until the early 1980s on the CompuServe network. Many of the PLATO games eventually turned up as commercial products (*Collapsar, Empire*, plus tank and flight simulators). In some cases, the PLATO games served as models for later commercial products.

In 1977, PLATO put on-line the first tank simulator (*Panzer PLATO*). This game was done for the U.S. Army Armor School at Fort Knox and was quite detailed and accurate. It was basically an improved version of the earlier (and much less detailed) *Panther PLATO*. *Panzer PLATO* was a very accurate simulation of armored vehicles. Even cannon shells were accurately modeled, so you fired smoke shells for sighting purposes.

Also in 1977, Chris Crawford published a tactical armor game (*TANKTICS*), which was redone and published for the Commodore PET PC in 1978 and thus became the first published wargame. *TANKTICS* was not in the same class with *Panzer PLATO*, but at least it could be played by anyone with a PET PC. *TANKTICS* finally appeared in an improved form as an Avalon Hill computer wargame in 1981. The game was not a big success. This was largely due to the limitations of the then-current "eight-bit" PCs (TRS80, Apple, Commodore). Poor graphics, low computing power, and only a few million PC owners made it difficult for any computer game to make a lot of money.

But publishers were encouraged, and from 1978 on there was a relative flood of wargame releases. This was brought on by the steadily increasing sales of PCs. It was at this time that publishers began to realize that science-fiction and fantasy games had greater sales potential than wargames. In the early years, the most popular computer games of these genres were the "adventure" games. These were interactive stories, without graphics, where you traveled around an unknown landscape (which you had to map on a piece of paper as you played) and asked the computer questions. These "text adventures" disappeared when better graphics became available in the mid-1980s and "graphic adventures" became the norm. The adventure-game format made the most of the PC's limited capabilities, and this style of gaming was never successfully applied to wargames. When better graphics capabilities became available, it was natural (and very successful) for the text-based adventures to move into graphic formats.

A year before the first PCs became available, Atari released the first game console, the Atari 2600. This 1976 bit of technology was basically a stripped-down PC, using an eight-bit microprocessor. It had no additional memory; that was in the cartridges. The 2600 didn't have a keyboard either, just a connector for a joystick and other simple game-control devices. The 2600 did have more powerful graphic routines than most PCs, because graphics were what these early arcade games were all about. The "arcade game" market grew enormously until 1981. In that year, Commodore came out with its very inexpensive PC, the C-64. In that year, Atari came out with an upgraded game console, the Atari 5200. Gamers weaned on the 2600 opted for PCs, and the C-64 outsold every other PC for the next few years. It wasn't until 1985, with the introduction of the Nintendo (with better software) that the arcade-game market recovered. The recovery took several years, but now the PCs were able to hold their own against the game consoles. The 16-bit game consoles introduced in 1990 were even capable of running PC-class vehicle-simulator wargames.

As the 1980s began, there were several striking computer-wargame designs. The earliest, and easily the most impressive, was *Eastern Front 1941*. Programmed by Chris Crawford on the Atari 800 PC, this game had outstanding graphics and featured the first "scrolling map." That is, the PC screen showed only a portion of the entire map, but by using the joystick, you could move this "window" around to show other parts of the map. The capability to do this was built into the Atari machines that, unlike Radio Shack or IBM, used processors (the computers' "brains") that had much greater graphics power. Although the Atari 800 series never became a great commercial success, it did demonstrate what could be done with the right hardware.

Before, and after, IBM arrived on the scene, game programmers performed minor (and sometimes major) miracles. Bruce Artwick brought out the first flight simulator in 1980. He even got it to run on the primitive TRS80 Mod I, an impressive programming feat by any standard. There were also several primitive tank simulators, generally with a science-fiction slant. Even then, publishers realized that the wargamers were a demanding lot. Do it right, or they won't buy it. The introduction of these first flight simulators, despite their primitive implementation, demonstrated how wargames would manage to compete with the more popular science-fiction and fantasy-type games. On average, a fantasy or science-fiction game will sell three to ten times as much as a wargame, except for the simulators. The only reason nonsimulator wargames were published at all and continued to be published was because they would sell enough to make a profit and, from time to time, there was one that sold very well. Another important factor was that a wargame did not require as much programming effort as a more graphics-intensive science-fiction or fantasy title.

This period, the early 1980s, was a watershed for computer wargames. While most of the potential customers were using Apple II or Commodore 64 PCs, and these machines were selling to millions of new PC users each year, the new IBM PC was beginning its inexorable climb to the top of the heap. Although the more powerful IBM machine had the potential to offer better graphics, this potential was not realized for over five years. Through the late 1980s, most PCs were thus stuck with primitive graphics capabilities. This was to the wargames' advantage, as a wargame did not need fancy graphics, or fancy anything else, like sound effects. An individual programmer could still put together a marketable wargame on his own, and they were mainly men doing the programming. So wargame programming, somewhat more than other types, became something of a labor of love for the creators. Low sales were kept viable by low costs. Wargamers should be more aware how important the underpaid programmers (themselves wargamers) were to keeping computer wargames on the market during the early and mid-1980s.

It took the computer wargames industry several years—indeed, most of the late 1980s—to fully digest the switch to the IBM standard. It was not a willing switch. Many, if not most, computer-wargame programmers favored the superior technology found in the Apple (Macintosh), Atari (ST) and Commodore (Amiga) PCs. But IBM had the numbers, and the Intel microprocessors that were the guts of the IBM-type PCs weren't all that bad. The programmers proved what the Intel chips could do, and produced ever-more-striking computer wargames from the late 1980s on.

Oddly enough, this march of technology produced ever more games on the IBM-type machines that had the same (easier to use) "look" long found on the Macintosh, ST, and Amiga machines. Indeed, the cutting edge of "ease-of-use" PC technology was to be found in game software. If a programmer wanted some ease-of-use ideas for his new word processor or accounting program, all he had to do was look at the latest games. That's precisely what a lot of nongame programmers did. After all, programmers were some of the most eager computer-game users. It was the perfect way to take a break without leaving the keyboard.

As the 1990s dawned, computer wargames were ready to take advantage of a number of new technologies.

## THE FUTURE ✱✱

The early 1990s will be best known for an enormous jump in computing power for the average PC user. But it's not just raw computing power that will change dramatically what computer wargames are, but all the things you do with that power.

Among the new wargame features possible are:

**Better Artificial Intelligence (AI)—** From the very beginning, PC-based wargames were meant to be played against the computer. The quality of the computer's play was dependent on the AI. While very good AI could be programmed into a small amount (about 5 percent or less) of scarce memory, more memory and greater speed allowed for a more formidable and less predictable computer opponent. The current crop of faster machines, with their much larger memory, allows for a full range of computer opponents, from pushover to formidable. In addition to different degrees of complexity, AI opponents can be given different personalities and even be made to assume the characteristics of notable historical commanders.

**Flexibility—** More speed and capacity allow computer games to deal with a lot more detail. Players are particularly fond of lots of reports and a playback feature that allows them to view how they played in a previous game session. The simulator games have, for several years, been able to record player actions and save the action to a file for later playback.

**Access to the Game's Guts—** Long missed by wargamers who remember fondly the access they had to what makes things happen in manual games, direct access to the inner workings is now possible with newer wargames. Computer-game designers are somewhat resistant to revealing exactly how things are done inside the game, as they have got accustomed to developing internal routines without the public looking over their shoulders (and second-guessing these efforts, rightly or wrongly). But the gamers are demanding access, and the ability to edit game parameters (probability tables and the like), and increasingly they are getting it. Note that this access is much more difficult than in manual games, as you can play a manual game incorrectly, and it will not "lock up" on you. A computer game is more sensitive to changes within the computer instructions, so it's a lot more work for the programmer to give the gamer access.

**Better Graphics—** The black hole of computing power is graphics. More than anything else, it is graphics that demand more and more powerful computers. When you consider that many games are approaching photographic-quality interactive graphics, it's time to realize that these graphics take up enormous amounts of disk space and require very fast machines to keep the action going at a realistic rate. Simulator wargames make the biggest demands on graphics and, as these are among the best-sellers, they set the standard. Other, less graphic-intensive types of computer wargame benefit by having more resources to play with. This generally lowers the price of nonsimulator wargames, which are also much cheaper to develop as they do not have to push, or even keep up with, the state of the art in graphics. But because of all the work done on simulator wargames, there are a lot more programming tools and experience out there for anyone to use.

**Hyptertext—** Increasingly, computer wargames have acknowledged their literary roots and have included a lot more background information in the games. This data is best (and most commonly) dealt with using a computer technique known as "hypertext." This technique is nothing more than the ability to hit a particular key when you are doing something in a game and instantly have a screenful of relevant information pop up. In other words, hypertext is context-sensitive information. Often, the screen of information has words or illustrations highlighted so that, if you want, you can bring up another screen of data on that particular item. Hypertext takes up a lot of computer resources (disk space and system memory) and only recently has there been enough space to accommodate it. Wargamers want them, so hypertext information systems are provided more and more often.

**Sound—** Next to graphics, realistic sound is the most wanted and most expensive to provide. We're not talking about sound from the shabby speaker on the IBM PC machines, but an additional component (a $150 board that slips into one of the slots inside the machine) that hooks up to headphones or high-fidelity speakers to give very realistic sound effects. Initially, these consisted of the sounds of the equipment and combat for simulators. By the early 1990s, the sounds included actual conversation. This consisted of messages from other (computer-generated) people on your side, the enemy, or a "computer adviser" telling you how you are screwing up. This last item is sort of spoken hypertext and is quite popular.

The next technical marvel on the horizon is the compact disc (CD). This is basically the same as your audio CD, but hooked up to a PC to play images as well as sound. However, the CD player for the computer is more expensive because computer access requires a sturdier and more precise device. Some of the PC CDs are available for under $500. The advantage of these items is that each computer CD can hold about 680 million characters of information. Sound and graphics take up nearly all this space. Ten seconds of sound can take up over a million characters of space, and one detailed graphic screen can consume half a million characters. The computerized role-playing games (RPGs) are thought to be the biggest market for CD games, as RPGs use a lot of graphics and sound. Wargames, even simulators, use much less, so wargames will be able to do much more with the enormous resources available in the upcoming CD-type games. Basically, you will see a lot more information in computer wargames using CD games. It was the desire for more information that created wargames in the first place, and the use of CD capacity will bring that aspect to undreamed-of heights.

This brave new world of computer wargames can be a little disorienting. Take, for example, the changing definition of "What is a wargame?" With paper wargames, there was little ambiguity. With computer wargames, the definition has been stretched a bit. This can best be seen in the product mix of stores and direct-mail vendors of wargames. Both stores and vendors present a list of products that they know, from experience, sell to the "wargame market." Not all of their products are wargames. In early 1992, for example, there were over 300 computer programs classified in the "wargames" category. However, nearly half the titles are simulators, most of them being aircraft simulators. The aircraft simulators alone account for the majority of the wargame sales. About half the titles available are what most wargamers would consider wargames. But the retailers know that wargamers will also heavily buy science-fiction and fantasy titles that are, basically, still wargames. There are no role-playing games included here, although many of them are more wargame than anything else. But fantasy games set up as conventional wargames do have heavy crossover sales. Especially when set in the medieval period, when many people did believe in magic, these games are bought (if not highly respected) by many wargamers. Science-fiction titles are another matter, as these games deal with a future that could be true, as opposed to fantasy realms that are neither past, present, or future reality to anyone except their most devoted practitioners.

Ironically, most computer-wargame titles are being sold by mail, just like earlier paper wargames. The software, electronics, and toy stores that provide most of the retail exposure can carry only so many games, and the additional wargames in print can still be profitably sold by the many direct-mail vendors. The major advantage of direct mail, aside from a much wider selection, is low prices. The average wargame costs $30 (in early 1992), with prices being 20 percent to 50 percent higher in stores.

Among the 45 percent of titles that are simulators, there are some curious subjects covered that are more a matter of wargaming branching out than nonwargaming moving in. These include titles like *SimCity, SimEarth, SimAnt, Powermonger*, and *Populous*. All of these simulate their subjects in great detail. Simulator products such as these have their origins in the complexity heretofore demanded (and tolerated) only by wargamers. While these games sell extremely well, wargamers make up a substantial proportion of their buyers.

Computer wargames have spawned some unique publishing angles because of the technology used. PCs become obsolete, and thus so do the games that run on them. The obsolescence is a result of continuing rapid developments in computer technology. The computers keep getting more powerful and cheaper. A minimal game system (48K RAM, 1 floppy disk, TV-type monitor) in 1980 cost $1,500. Adjusting for inflation, it costs only $1,000 1980 dollars to get a 2,000K RAM, floppy and hard-disk and high-resolution color monitor in 1992. For the last 10 years, basic computer-game technology has been turning over completely once every three or four years. While popular paper games could be kept in print for decades (I'm still collecting royalties on *PanzerBlitz*, a manual wargame published in 1970), no computer game outlives the hardware it was designed to run on. Actually, most computer wargames last only a few months in the stores, although they hang on longer from mail-order vendors. But many of the computer-wargame best-sellers are being redone for each new generation of hardware. Since the late 1980s, more than half of the best-selling wargames have been redone to work on the latest hardware. This has been most common with the simulators, which are the most hardware-intensive computer games.

And then there are the add-ons. Because computer games are basically huge computational engines, if you load new data into most games, you have, in effect, a new game. It was the aircraft simulators that first discovered this angle. Microsoft's *Flight Simulator* had scenery for the player to fly around in, and after a few generations of this mid-1980s design, the scenery was realistic enough to be based on actual places. Soon players began clamoring for scenery from their part of the country, and eventually additional "scenery disks" were published to accommodate this demand. Most flight simulators involve fighting other aircraft or ground targets, so it became common to publish additional disks with new battle information (enemy aircraft or target areas). This was particularly the case during the 1991 Iraq war, where several combat-flight simulators quickly came out with Iraq-war scenario disks.

The publishers were quick to realize that these add-on disks were much cheaper to create than the original game (10–20 percent of the original cost), yet could be sold for about half the cost of the original game. Nothing like a little financial incentive to spur things on. The gamers were quite happy with this arrangement, as the new material extended the play value of a favorite game. Moreover, a common add-on was a ''scenario editor'' that allowed gamers to build their own additional features for the game. Still, most gamers preferred to let the publisher develop the new information. Another aspect of this is that the prospect of more profitable add-on disk sales makes it possible to risk more money in more elaborate games. The add-on kits are a growing trend, with over 5 percent of the items currently (in early 1992) found in a typical software store being add-ons.

Remainder sets are another development of the 1990s. Because stores can return unsold games to the publisher (''return privileges''), there is a lot of old computer-wargames inventory lying about. In the last two years, many publishers have sold off this old inventory (''remaindering''; before that a lot used to be destroyed) to ''remaindering'' outfits that now box three of the old games in one box and sell the set at a low price. The publishers often get 10 percent or less of the original price per game when they remainder. The idea seems to be working, as you see more and more of these ''remainder sets'' in the stores. Apparently, the publishers don't think this cannibalizes the sales of their newer games. It looks as though this may become a permanent feature of the computer games market. It's great for people with older (less powerful) PCs, as the newer games often demand more computing power than they have. This is probably why publishers don't mind this new form of ''competition'' from their older, much lower priced, games.

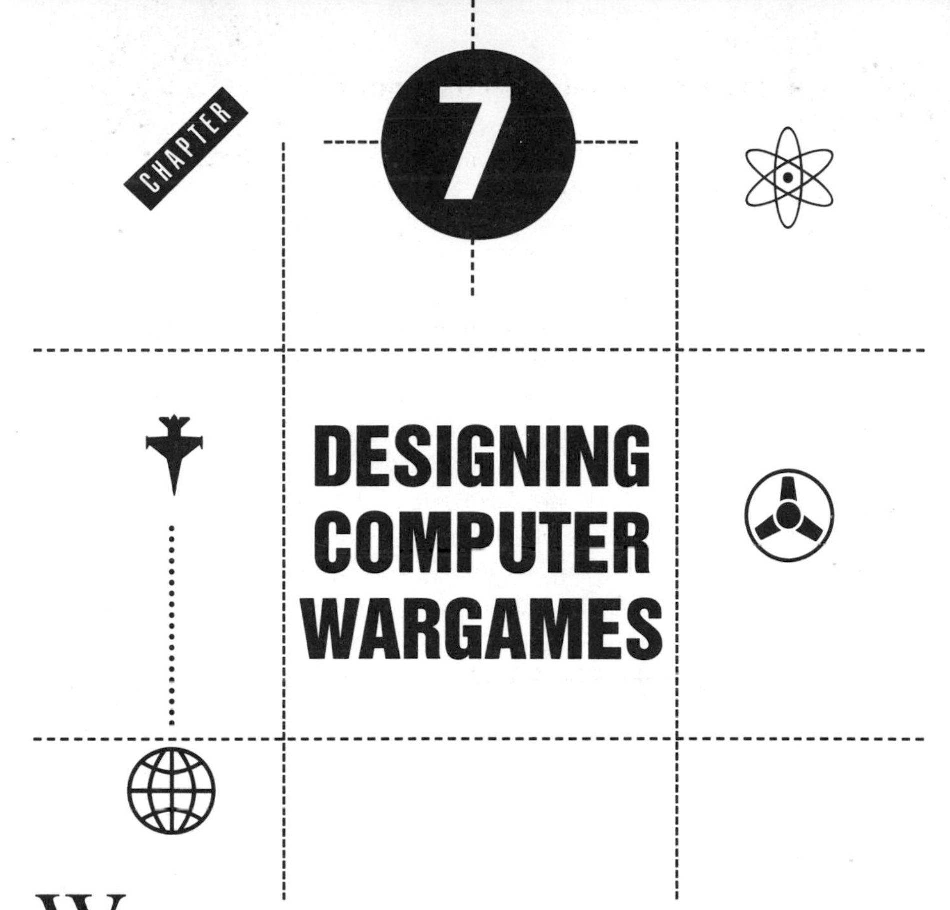

# CHAPTER 7

# DESIGNING COMPUTER WARGAMES

Wargames on computer have had a mixed record. The "medium" (the computer) is substantially different from the paper and cardboard used in manual games. What works on paper does not always work so well on the computer. The computer has capabilities that are often not exploited in computerized wargames. Those wargames that do tend to emphasize the computer's strengths tend to be situations like vehicle simulations. This makes sense, as a computer can handily deal with the numerous calculations needed to simulate aircraft, ship, or AFV (Armored Fighting Vehicle, tanks and the like) activity. Manual games have a lot of problems with these situations, so computerized vehicle simulations have done very well, while their manual counterparts languish unplayed by gamers with increasingly less time for games.

Computerized versions of nonsimulator wargames still constitute an evolving area. Programmers have been designing computer wargames for over 20 years, but game designers have not been working on them for nearly as long. This has always been a major problem with a lot of computer wargames, that they were (and largely still are) designed by programmers who are often new to game design. The problem lies in the fact that game design and programming are two different skills with only some conceptual overlap. Though a lot of manual wargamers are programmers, only a few programmers start out as game designers. This has produced some awful computer wargames. While the cost to the players of these wretched games has been high (more in terms of aggravation than money), the experience has enabled those programmers with a talent for game design to rise above the pack. Chris Crawford and Gary Grigsby are two of the most notable of this new generation of designers, and there are a dozen or so who are nearly as good as these two. (I won't name names, lest I inadvertently forget someone who deserves to be mentioned. You know who you are, including the foreigners.) But there are still a lot of neophyte game designers who may be great (or sometimes dreadful) programmers turning out painfully inadequate games. This chapter is for you.

Thus the first thing you must consider when creating a computer wargame is whether you are going to do both the designing and programming, or only one of these chores. This is not a trivial decision. A superb programmer may be a mediocre (or simply lackadaisical) game designer and will produce, at best, a good-looking game that has little useful play (or historical) value. A good game designer who is an inadequate programmer won't get very far either. The third solution, to have a good game designer work with one or more good programmers, usually produces good results once the communications problems can be overcome. This team approach is becoming increasingly common, particularly since current computer wargames are so elaborate that many different specialists must be called in (for programming, graphics, interface, sound, and documentation). Adding a game designer to the list is no burden and is usually a decided plus. The last solution, having someone who combines the talents of a programmer and game designer, is rare because that combination is extraordinary.

Before we get into the nuts and bolts of this, I ought to present my own credentials. This will put what I say here in perspective. I've designed five computer wargames. For three of them, I did the design and someone else did the programming. The first one I programmed myself. It was an experiment, which was just as well. It didn't turn out that badly, but it was far from publishable. It was called *Rus* and was based on the Viking invasion of Russia in the ninth and tenth centuries. I programmed it in Microsoft BASIC (on a TRS80 Mod I) in 1981, so that I could get an idea what programming computer wargames was all about. I had some background in programming, having learned the rudiments of COBOL and RPG II in the 1970s, in addition to the primitive language of HP's first series of programmable calculators. But COBOL and RPG II were dreadful languages, and although I had access to a minicomputer running those two languages, I did little with my programming knowledge beyond reading the program listings of the programmers that worked for me. I had done several business applications in BASIC, a language I learned in 1978.

The Rus (as the Russians called the Vikings) advanced down Russian rivers in the eighth century A.D., plundering and settling as they went. So in my game, you proceeded down one of the Russian rivers, encountering (and dealing with) all sorts of situations as you went. It was a revealing experience. I quickly learned that my game-design ideas rapidly outstripped my programming skills. I was comfortable in BASIC, and I knew what peeks and pokes could do within the operating system. After finishing the game, I decided to leave programming to those with a knack for it.

The second game I programmed was done on a dare. What brought that about was the appearance of the Lotus 123 spreadsheet program in 1983. I had been using the earlier VisiCalc spreadsheet to great effect since late 1980, but 123 was a supercharged VisiCalc with a macro language. The macro language was, in essence, brain-damaged BASIC. I did a lot with macros, and still do. On a dare, I created a wargame on a spreadsheet. Actually, the first spreadsheet wargame was done on the CP/M version of Microsoft's MultiPlan spreadsheet. I ended up doing versions of this wargame on SuperCalc, Symphony, and Quattro. Someone else got it going on the Excel spreadsheet program. I began giving it away in 1983, and that "wargame" played a role in getting the military to use spreadsheets for combat modeling. This type of computer wargame is not slick enough to be a commercial product, but it gets a lot of real wargaming work done.

My third computer wargame was more polished, and recognizable. In 1985 I was asked by an old army friend (Ray Macedonia, recently retired from running the Wargames Department of the Army War College) to create a manual wargame on tactical armored warfare. He needed it for the work he was doing with his new employer (AVCO, later part of Textron) on antitank weapons. So I created the manual game in a few months. The AVCO people liked it so much that they asked to have it turned into a computer wargame. They gave me a Symbolics workstation, two programmers, and a lot of money; three months later we had it up and running. Neat game, full-color graphics, AI, and everything.

While I had never done a computer wargame like that before, I already knew how to spec out (writing the specifications for) a project for programmers, as I had been doing that since the early 1970s and through the 1980s as a supervisor for teams of programmers doing financial-modeling programs. In 1989 I got involved with the GEnie computer network and ended up agreeing to design a multi-player game (over 300 players) of the Hundred Years' War. The programming was done on a mainframe computer, and all the players got together via their PCs and modems. That game went into alpha-testing in 1991 and went on-line for paying customers in 1992. In 1991 I was approached by 360 Pacific (publisher of many computer wargames, including the best-seller *Harpoon*) about doing a computer wargame on the naval war in the Pacific. I agreed, and the spec was done by October 1991, with programming to continue through 1992.

It is from this perspective that I will describe how one should go about designing a computer wargame. Some of the material presented here is from the lectures I have been giving to military wargame designers for the past dozen years. That will just broaden your knowledge of designing computer wargames a bit more. Note that we are talking here about designing, not programming, a computer wargame. Actually writing the program code for a computer wargame is a whole different matter. The programming techniques are not only very arcane (and largely understandable only to programmers) but change substantially from year to year as computers and programming tools become more powerful and, well, different. Designing computer wargames consists of using principles and techniques that are less subject to change.

## THE SPEC ✸✸

Keep in mind that a computer does what you tell it to do, not what you want it to do. Unlike people (some people, anyway), you can't just tell a computer what you want done and expect your request to be carried out. Computers require explicit instructions. These are called computer programs, or computer software. The terms "program" and "software" are often used interchangeably.

Programmers make their living by turning someone else's need for a computer to do something into a program that will make that computer do it. The programmer, however, requires precise (or at least unambiguous) directions from the user of the program. In this case, the user giving direction is the game designer and these directions are delivered in the form of a "specification" or "spec."

Specs come in many different forms. On one extreme, you can get many pages of flowcharts (boxes of different shapes connected by solid and dotted lines) showing which data goes where and does what. These flowcharts are accompanied by equally voluminous instructions on how to set up data files (lists of information), what formulas to use, and how the processed data should look when it is presented to the user (on the computer screen or printer). On the other extreme, you may get some verbal instructions accompanied by a few notes scribbled on a piece of paper. My approach to a spec is somewhere in the middle, as I have found that the flowchart approach gets ignored (at least a lot of detail gets ignored) by the programmers, while the other extreme leaves the programmer confused and inclined to invent whatever he needs and hope that it's what the user meant.

Naturally, I'm going to push my own personal version of what I think a spec should be, if only because it's a system I have used successfully for many years with dozens of different programmers. I count my approach a success because most of the programs worked well and most of the programmers are either still working for me or at least return my phone calls. What we are describing here is a spec for a computer wargame, any computer wargame, and my spec is divided into three parts: input/output, data bases, and internal-model procedures.

First comes the *input/output* (I/O). In plain English, this means what the user sees on the screen (output) and what commands the user can give to the program to obtain a desired result (input). This is often called the "look and feel" of a program. The easiest way to prepare this is to do a mock-up on the computer screen. These days, one of the many "paint" programs is used to literally draw ("paint") the various proposed screens. Some years ago, I used a program called Demo (clever name, eh?) to build screens. The advantage the Demo program had was that you could make the controls on the screen active, to bring up additional screens and similar things. I stopped using Demo when I realized that for my purposes (my programmers didn't need *that* much direction), a much simpler method would suffice. So I just typed up what the input and output screens would look like. Games use very complex screens, often with lifelike terrain, vehicles, or people on them. But these don't have to be shown realistically, and a simple notation will do. The important thing is to let the programmer know what will generally go where on the screen.

When you think about it (and it's a good idea to think while speccing a program), there are not that many screens involved in most games. There might be a lot of different graphics put on the screen, but this is just popping a new picture on to the same screen. Yes, that's one of the secrets of computer-wargame design, to move and vary a small number of items to make it look like a whole lot. Just like real life.

The foundations of any automated model are its *data bases,* and they form the second important part of the spec. Computer wargames use quite a few different types of data bases, more or less in the same way manual games use tables and charts of information. These include the following:

• **Terrain.** The terrain data base can safely follow the conventions used in the manual model, even to the extent of using the same cell structure and terrain information used in each cell. There is rarely any justification for the extreme detail found in many digitized terrain systems.

• **Order of Battle.** List of units to be used in the model and the capabilities they need for interaction with the other elements of the model. Unlike with a manual wargame, you can safely use a lot more data, because the computer has a lot of capacity to crunch numbers without even breathing hard.

• **Attrition Tables.** What happens to weapons and targets when weapons are used. These are the Combat Results Tables (CRTs). Computer games can use more complex and multiple CRTs.

• **Terrain Effects Tables.** Effects of terrain on units. Again, effects can be more complex and detailed than in manual games.

• **Victory Conditions Tables.** Goals of units or groups of units.

• **Supply Tables.** Supply holdings and consumption for various units.

• **Doctrine (AI) Tables.** Artificial Intelligence Capability. Rules for unit behavior. SOPs for combat and movement. Tables for command control and panic for various units. Leadership behavior, etc.

• **Other Items.** Any other items that the user feels are needed for the model.

• **Scenario Tables.** These specify which of the above items are to be combined into a particular battle situation. Preferably this is kept in a library (on the computer's hard disk) that the user has easy access to with an editor and report generator.

*Internal Model Procedures* are otherwise known as algorithms in computerspeak. The manual model does not translate directly to an automated version. Some adjustments and design modifications must be used. Typical internal model procedures include:

• **Positioning on Terrain.** These generally use an *x, y* coordinates system. It is easier to implement on a computer and will give you roughly the same accuracy as the hex-based manual models.

• **Combat Activity.** This follows a Sequence of Play very similar to that used in the manual model. The higher speed and automatic operation of a computerized wargame combine to give the appearance of simultaneous activity. Combat models perform the following operations which generally take place in alternating fractions of a second (for each phase in the Sequence of Play):

**1.** Detection, or using sensors to determine if enemy forces are present and/or within range of weapons.

**2.** Checking for Damage from the enemy from operational causes. If units are damaged, the system produces physical effects (Degradation of the units' performance) and often psychological effects (Modification troops' will to continue operations).

**3.** Fire or Flee, in which the system analyzes the situation. If logic, or psychological effects, dictate, the system withdraws the units from combat. Otherwise, combat continues.

**4.** Recording actions in the "diary" data base for later report generation.

**5.** Repeat the steps above, continuing until end of time period allotted for operations, or until victory conditions specify ending operations.

- **Victory Calculation.** This is done continuously and with great accuracy, although the user may want to see the probability of victory at certain points in the run.
- **Scenario Deployment.** Much more rapid than in the manual model. A good user interface allows the player to rapidly explore many aspects of the game with great ease of use and accuracy.
- **Scenario Generation.** Using an electronic library of scenario building blocks (maps, unit organizations, doctrinal SOPs, etc.) the user can quickly build new scenarios or edit existing ones. The edit function also allows building blocks to be created or edited.
- **User Reports.** These can also be edited as to format by the user. The principal items a user would generally require in a report would include:

**1.** *OB Status*. Starting and finishing status of units.

**2.** *Terrain View*. Which terrain configuration was used (large/small scale via zoom).

**3.** *Victory Status*. What the player is allowed to know, given intelligence constraints and other factors, about how close either side is to victory.

**4.** *I/O Routines*. How to handle interaction between model and user. The user must be able to easily modify the following program instructions. This is done by setting up the routines in table format and programming to prevent the user from doing anything that would crash the system. If the user unintentionally does anything silly, the system will simply display these actions on the screen, allowing changes to be made and their results observed until the user is satisfied.

**5.** *Movement*. A combination of historical movement capabilities (based on field tests and experience); SOPs (standard operating procedure) for movement, right out of the book; and limiting factors, primarily psychological, that don't make it into the field manuals or peacetime soldiers' knowledge.

**6.** *Combat*. Similar to movement. In this case it applies to the servicing of weapons.

**7.** *Intelligence*. How much information of its own and enemy forces does each side have at start and during each intelligence-cycle time period. This time period is based on how long it historically took for new information to reach the commander and be acted upon.

**8.** *Other*. As needed.

## USER DOCUMENTATION ✱✱

In addition to the main parts of the spec I've just described, user documentation, growing in popularity, is also part of your spec. Various kinds of documentation include:

- **The Players' Manual.** The instructions for actually playing the game should be imbedded within the game via a "help" system. Printed instructions should do little more than list which keys do what or give an overview of the menu system. Help should be context sensitive (when you invoke Help, it first brings up a Help screen relevant to where the player is in the game.)
- **Hypertext Historical System.** It's increasingly popular to have a context-sensitive historical manual built into the system. When invoked, the system brings up historical data relevant to where the player is in the game. The "hypertext" refers to a system whereby key words on the screen are highlighted to serve as a command (when the cursor is placed over them and activated) to pop up another screen relevant to that key word.

## QUALITY CONTROL ✱✱

- **Testing.** Keystroke emulators are useful to automatically run the system through an extensive suite of hypothetical user operations. Some of these test programs will also capture the screens and other results to a file where this test data will be compared to what should have happened, and then flag error conditions.
- **Double Team.** A system I often use, which is expensive, is to have two teams of programmers overlapping (and even duplicating) each other on a project. This greatly reduces the chances of the project being late, or sloppy. I doubt if game publishers can ever afford this, but I thought it worth mentioning. It works.
- **Compile the Source Yourself.** Learn to use a code analyzer, etc. This is a programming function, but if you are overseeing a project, it's a good way to keep on top of things. You can make changes to data in the program (to correct something), recompile, and see if that fixes it. Only programmers should change code, and sometimes even changing data will cause a program to fail.

## ARTIFICIAL INTELLIGENCE (AI) ✱ ✱

These are the routines that enable the computer to operate as a worthy opponent for a human player. These AI routines are not as complicated as they might appear, and need not take up a lot of space in the program. It is possible to get powerful AI routines into as little as 2,000 bytes (''characters,'' or ''2K'' in computerspeak).

AI is driven by doctrine, i.e., the standard procedures that armed forces use when in combat. Every armed force has a doctrine. Some doctrines are more efficient than others, but that's something you have to dig up in your research. There is also the problem of ''theory versus practice'' in doctrine. What the doctrine says is not always what the troops do. You have to do more research to find out how the troops deviated from their doctrine, to what extent, and how often.

For the commander, combat is a rather simple process: at least as far as decision making goes. There is often imperfect information. That is, the commander is often not sure of the status of his own troops and is even less well informed about the enemy. Thus the commander usually makes simple decisions. To mix things up and keep the human player on his toes, you should also use a random selection from two or more possible moves by the AI side.

AI becomes complex when you want it to measure its side's situation against the human player's. This routine is driven by the victory conditions. The measurement is a combination of the combat strength of friendly and enemy units, the ''value'' of their current position, and the ''value'' of enemy positions. Normally, the AI would select the highest-value enemy positions to go after, but not always. By randomly selecting one of the most valuable enemy positions, the AI player gives the human player the impression that some real thought is going on. If nothing else, it makes the AI side unpredictable, and thus dangerous.

Creating the AI routines is a game in itself. You'll simply have to practice. It's not so hard. I did many successful manual AI systems (for paper wargames) before I did my first computerized systems. With the experience gained on the paper games, I had no trouble at all with the computerized AI. Your mileage may differ.

## THE COMPUTER WARGAME DEVELOPMENT TEAM ✱ ✱

First, there's the designer. Actually, before the designer there's the publisher, who has to agree to put up the money and distribute the final product. The publisher generally gets no respect and less recognition. While I've designed over 100 games, I've published nearly 400. Designing is generally fun, publishing is customarily hard. That said, once the designer has delivered a spec that makes some kind of sense to the programmer, the programmer has to turn it into a computer program, the software, the game you can play.

In my experience, it is best to use one programmer, plus support staff. The more programmers are used, the more time is wasted in programmers keeping tabs on each other. Modular programming is not practical, as with this approach much of the system design is done during code development. Programmers must be diligent, willing to work six to seven days a week, leap tall buildings at a single bound, etc. The programmer support staff should include:

- **Designer.** Whoever designed the manual game, or drew up the spec without a full-blown manual game. The designer's job is to ensure that the programmer doesn't get lost while implementing the spec. It's best that the designer be kept in the process as the programming goes forward. There's always the chance that the programmer may try to turn the game into his design (either on purpose or by accident), a development that rarely works out very well.
- **Development System Experts.** To help the programmer with quick solutions to technical programming problems. These people give advice when asked, they don't write code. In *Hundred Years War,* this consisted of the folks who ran the GEnie mainframe computers and the worldwide communications system over which players would connect with each other as they played the game. In *Victory at Sea,* the lead expert was Gordon Walton, who led the earlier *Harpoon* project for the publisher, had designed and programmed some wargames himself, and could solve a lot of problems by simply pointing where they were and what the easiest solution was.
- **Chief Testing User.** Keeps everyone honest by testing the system every step of the way. Does nothing but test and make a general pest of himself. Sometimes the designer takes on this task, sometimes it's someone working out of the publisher's offices. Generally, this person is in charge of testing and keeps track of bugs as they appear and checks to see that they are taken care of. A currently popular title for this job is "quality control manager." Sounds better than "chief pest."
- **Researcher.** Digs up additional operational data not already present, but often implicit, in the manual model. The computer wargame can handle more detail, and it is often useful for it to do so. In both *Hundred Years War* and *Victory at Sea,* Al Nofi was the principal researcher.
- **Programmer Assistant.** Handles routine tasks for programmer

(common data-entry tasks, system maintenance, ordering the pizza and Jolt cola, doing the paperwork, etc.).

• **Project Manager/Expediter.** Keeps management at bay, solves project-related problems quickly. Hand holder, ass kicker, cheerleader, etc. This is usually a management representative who reports directly to the publisher. Often has to yell at and argue with publisher also. For some reason, I like this particular job.

In the commercial computer-wargames industry it is currently fashionable to use job titles borrowed from the movie business. Most of these wargames companies are in California, although largely northern California. Lifestyle envy? Anyway, the head of a computer game project is the "producer," the chief programmer is the "director." I have not confirmed this, but the sound specialists may now be called "Foley editors." Given that all the "actors" are electronic, where does that leave the casting couch?

## HARDWARE FOR COMPUTER WARGAMES ✱✱

• **IBM PC (and Clones).** First introduced in 1981, this machine does not have the best combination of features, but it can do the job. Uses the Intel 16-bit 8088 or 8086 processor. A wide array of add-on components make the IBM PC series the most flexible machine available. It is widely available—there are over 60 million installed worldwide. The standard machine in most organizations, with the widest selection of software. Most major new software first comes out in an IBM PC version. The original (1981) IBM PC had twice the throughput of older eight-bit machines (Apple, CP/M, Commodore 64, etc.). This is now considered a low-end machine, but it can be boosted much higher with add-ons. About half the PCs in the world are of this type. I still have a 1983 Compaq (a portable clone of the original IBM PC), which I keep as a spare. I've added a hard disk and a high-density disk drive. Nine years old and still going. Gets turned on four or five times a year and still works.

• **IBM AT Type.** Introduced in 1984. Also known as the ISA (Industry Standard Architecture) type. It has three to ten times the throughput of the original PC and can be boosted even higher with add-ons. Based on the Intel 16-bit 80286 microprocessor, this was the baseline machine for many new computer wargames in 1991.

• **386 Machines.** Introduced in 1986. Based on the Intel i386 32-bit microprocessor, these machines have 15–30 times the speed of an original PC. The standard machine of the early 1990s.

• **486 Machines.** Introduced in 1989. Thirty to 70 times the speed of an original PC. Based on the Intel i486 32-bit microprocessor. The standard machine of the mid-1990s.

• **586 Machines.** To be introduced in 1993, with 60 to over 100 times the speed of an original PC. This machine will also have much more powerful video capabilities, enabling it to create movies ("full action video") on the computer screen. Based on the Intel i586 32-bit microprocessor, this will be the standard machine of the mid- to late 1990s.

• **The Clones.** These offer the same capabilities as IBM models at lower prices. Some clone models also outperform IBM machines at a lower price. Compaq was the first major clone, in 1983. Others followed, pushing prices lower and performance higher each year. Major clone-maker PCs considered as reliable (if not more so) than IBM.

• **Macintosh.** The current Mac series is equal to the 386/486 and high-end AT machines, but costs a third more. Graphics about the same as IBM EGA and VGA modes. Original Mac and Mac SE a cut above XT in speed, but more expensive and not numerous enough to support widely distributed software. Later Mac models matched Intel microprocessor machines (IBM and clones). Many games running on IBM machines are not available for the Mac.

• **Low-Cost Machines.** *Apple II* and *Commodore 64/128* have limited power, but are still widely available. There are no longer any new games for Apple II, and only some for the Commodore. *Amiga*—Excellent hardware capabilities, but not a major factor in the U.S. market (most sales are in Europe). Good selection of software, but increasingly distant third after the IBM and Mac. *Atari ST*—Similar to the Amiga, but not as technically advanced. Also sold widely in Europe, but less and less in the United States. The ST may not survive long in the U.S. market.

To give you some more insight on how computer wargames are designed, I include here some material on the two computer wargames I was working on as this book was being written.

## VICTORY AT SEA ✱✱

This is a game on naval warfare in the Pacific during World War II. In effect, it's a strategic game of the entire war in the Pacific with the option to drop down at any time and fight it out ship to ship (or, more likely, aircraft to ship). It's the first of a series of games. Additional disks will be published later to cover other theaters of the World War II naval war.

Below is an extract from the introduction to the *Victory at Sea* spec, delivered on October 15, 1991.

## NOTES ON GAME COMPONENTS

### THE MAP DISPLAYS

At the moment, you have a list of ports and instructions to scan any Pacific region map into your paint program. Then implement latitude and longitude coordinates to overlay the ports list onto this map. Because a degree of longitude varies as the cosine of latitude, the length of a degree of latitude varies from 69 miles at the equator to zero at the poles. We don't go quite that far, but do reach two thirds of the way to the North Pole (60 degrees north), meaning the degree of latitude is under 50 miles that far up. The map will have to be adjusted. That should be no problem as you will have the 140 principal locations listed in terms of latitude and longitude. We can supply more as needed, although we don't want to crowd the map. We counted over 500 atolls and islands capable of supporting some kind of garrison (many at great expense in shipping, as even water would have to be brought in to some of the drier ones). Some of these were actually garrisoned by the Japanese (Tarawa). Once you have the map sorted out electronically, send it to me, and we'll indicate the "atoll clusters." I thought it better to leave that complication for last. It will make some programmer's life easier.

### INTERFACE

The interface Gordon and I worked out was earlier keyed to routines in the game. This has to be updated, and I will do this in the next week. There's more than enough stuff for each existing screen mock-up.

### DATA BASES

The data bases are complete, although we'll update them right down to the wire as we uncover new stuff. Nothing radical, but there's always something. As all the data bases are in spreadsheet files, we can put them in any form you want for uploading into the source. I do that a lot. I don't like it, but I've got pretty good at it.

### ALGORITHMS

Over 95 percent of the needed algorithms are done. Some of the key ones have been tested and proofed using a Monte Carlo system I've developed. However, it requires a bit of coding and screwing around to get them into a format where the Monte Carlo testing will work. You know what Monte Carlo is? If not, it's a model that tests an algorithm or, more often in my case, a series of algorithms over thousands of iterations, and gives a reliable picture of what the system will do over many iterations. My system produces graphs as well as summary stats. I can then go back and tweak it until I get it to do what needs to be done. I will continue to test the game algorithms this way until they are cast in code, at which time we should Monte Carlo them again. The naval surface-combat routines have gone through over 6,000 iterations like that. But many of the routines will not be fully testable until they are in code. The systems are set up so that I have a lot of numbers I can change to bring the system into the proper balance. Once you start writing code, give me initial executables with as much data in external tables so that I can tweak. I can use a hex editor to change ASCII data in a binary executable or overlay. Changing binary directly is a bitch. Could also use some Monte Carlo modeling capability, mainly the ability to have *x* number of iterations of a situation with both sides on AI and loss stats sent to an ASCII file in tabular format for analysis.

### ARTIFICIAL INTELLIGENCE

The keys to realistic-combat AI are hidden information and randomness. If you have that, you can employ quite simple routines. As Clausewitz and Sun Tzu point out, even simple plans are difficult to execute in combat. The Artificial Intelligence data is presented in narrative form with the probabilities embedded in the description of the routines. I'm still having some problem with the strategic AI, but nothing that a little more application won't solve.

### GENERAL ADVISORY

This is the first time in nearly 20 years that I've had to deliver a spec to programmers I've never met. I usually do the programmer hiring, so I know what they are capable of and what they prefer to get in the form of a spec. It also helps that my programmers know I sign the payroll and bonus checks. In any event, I'm delivering the AI stuff in a pretty generic format so that the programmers can get a good sense of what it is going to do and then get back to me with whatever alternative format they might prefer. It won't take long to reformat it once I know the preferred format.

### FOOTNOTE FEATURE

We can cobble those together from notes, but for a full-blown version we'll need a codicil on the contract to clarify the book contract Al and I are now negotiating with my publisher. The big version would, in effect, be a book on a disk and would be a great selling point that would add little to the production costs (an extra disk) and nothing to development (the bits that are already going into the FootNote feature would simply be larger). Tom says he's amenable to signing the agreement needed to protect the book copyright.

**COMMUNICATION**

Until you figure out exactly how you're going to code this thing, you'll have a lot of queries. Since financial and military models don't mix, you won't be able to get hold of me when I'm downtown (material here deleted as I don't want too many people to have an easy time tracking me down when I'm trying to get work done). Otherwise, leave a message on the answering machine. I will E-mail confirmations on all phone discussions of spec changes, but do it via E-mail as much as possible. Normally, I check E-mail twice a day: between 4:00 and 8:00 A.M. and 4:00 and 7:00 P.M. Eastern time (1:00 to 5:00 A.M. and 1:00 and 4:00 P.M. Western time). If you alert me that you will be throwing a lot of queries at me, I can usually arrange to check mail between 11:00 A.M. and 2:00 P.M. Eastern time (8:00–11:00 A.M. Western time). In most cases, I take care of E-mail queries immediately (especially when I'm at home, where all the computer records and printed research material is available). Thus, in over 80 percent of the cases, you will have an answer in under six hours. Those I can't answer immediately I usually pass on to Al Nofi (ANOFI on GEnie), and he usually gets back in 24 hours. I keep copies of all incoming and outgoing E-mail and keep it indexed, so I can find anything within seconds. One helluva organized sumbitch, ain't I?

**SPEC FILES**

Files (with one exception, two versions of each, .WK1 is 123 spreadsheet version and .ASC is IBM ASCII version with CR/LF at the end of each 72 character line):

- **VASSPEC.NN.** This note.
- **VASGAME.** Game procedures. The game rules. Has a functioning menu system if you have a spreadsheet compatible with 123 macros (I used Quattro Pro in 123 mode).
- **VASSHIPS.** The ship stats. This was larger until we took out the German, Italian, and most of the French and British ships (the latter two nations did have some ships in the Pacific). The game information is on the left-hand side. To the right is historical information, some of which is used to calculate game values so it has to stay in place.
- **VASDIV.** The ground and air units.
- **VASPORTS.** The map information as well as capabilities of ports and bases.
- **VASAIR.** Not really needed by the codemakers, just us showing off. We had to compile this one in order to come up with a dozen or so numbers for "average" aircraft stats by year for each side.
- **VASGUYS.** The leaders, and all their stats.

• **Implementation Sequence** (as I see it)

Strategic Display
Unit List
Strategic Movement
Tactical Display
Tactical Movement and Combat
Strategic Combat
Strategic AI
Tactical AI
Leadership
Other

• **Program Sequence**

Load main executable
Player chooses scenario option
Program initializes scenario and begins play
Player input
Change production
Reroute supply
Movement orders

• **Computer Procedures**

Strategic Search
Tactical Search
Strategic Move
Tactical Move
Strategic (Aggregated) Combat
Tactical (detailed) Combat
Logistics
Strategy Selection
Tactics Selection
Etc.

I used the above to guide my priorities in conditioning the enclosed material for delivery. Let me know how it's supposed to be so I can shift gears accordingly.

• **Resolution**

One-Day Strategic
One-Hour Operational
30-Sec Tactical

A footnote on this project. While the official start date was July 15, it was largely delayed for six weeks because a few days after signing the contract, my copilot, Al Nofi, had to have emergency eye surgery. His eyesight survived, but it was not until early September that he could get into the research big time. You always have to make allowances for disasters.

### *HUNDRED YEARS WAR* ✱✱

As this book was being written, the game *Hundred Years War* is in alpha-testing. This game represents one aspect of computerized wargames that is just now coming into its own, the use of modems for multi-player games via telephone-linked computers. For nearly 10 years, computer networks such as CompuServe and GEnie have offered multi-player games, but none of these have been history-based (or, at best, only vaguely so). To remedy this situation, I signed a contract with GE to design a multi-player game (300 players) of the *Hundred Years War* to be played on its GEnie system. Al Nofi (research) and Dan Masterson (programming) comprise the rest of the team that is re-creating 14th and 15th-century England and France (plus Italy, Spain, parts of Germany, Scotland, Ireland, and sundry other adjacent areas). The game will cover economics, religion, and politics in addition to the purely military aspects.

As a player, your objective is to ensure the growth, prosperity and survival of your family line (each day of real time equals three months of game time). You may start out the game as anything from an impoverished Gascon noble to a mighty earl of England. Your degree of victory is rated on how much you increase what you started with during the century or so it takes England and France to settle their dynastic, military, and economic differences. The game allows a player to operate on two levels, either as a free-wheeling adventurer, living only for battle, tournaments, or the hunt (a rather common attitude in those times), and/or as an ambitious and able administrator of estates and participant in the affairs of state. The latter course is more rewarding in the long run, but the former can be more fun for the mash-and-bash set. There's a little of that in all of us.

In this game, the computer is used in areas where it does the most good. The medieval economic system is run by the computer, as medieval economies were fairly complex. Although most of the population lived by farming, about 10 percent produced a variety of manufactured goods and traded them over wide areas. There was money about, a bureaucracy, and a heavily armed nobility. At any given time, at least 1–2 percent of the population was in arms. Most of these troops were mercenaries and either they were paid or you suffered their depredations as brigands. Many mercenaries turned to brigandage once large armies were demobilized, creat-

VASIF05

STRATEGIC MAP VIEW

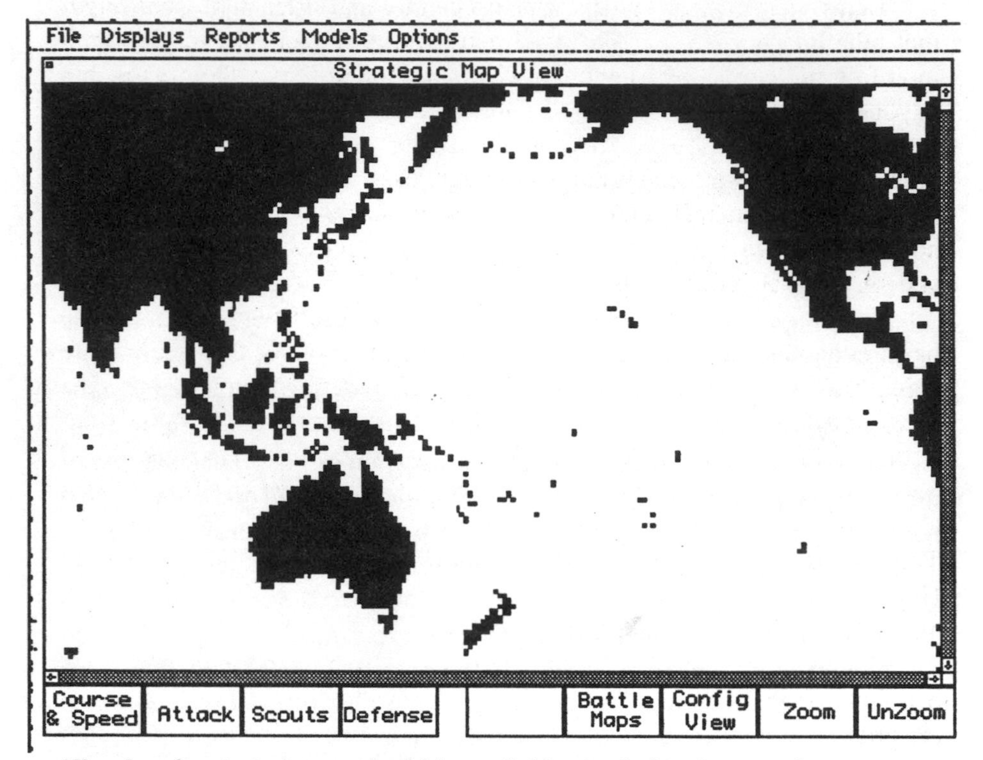


**All orders for strategic map should be available attached to the map view.**

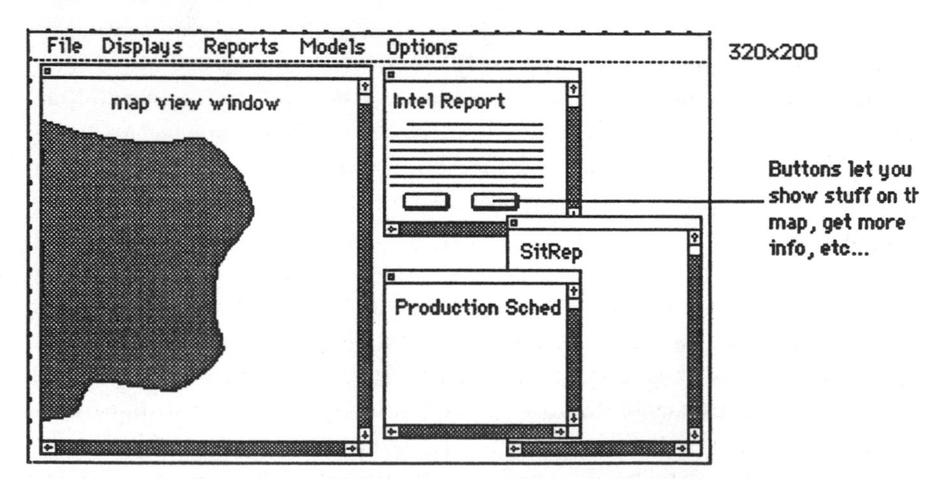


**Assumption is resolutions of 320 × 200, 640 × 350 and 640 × 480 are supported in the game (also 512 × 342, 512 × 384, 640 × 480 and large for Mac)**

ing a constant state of conflict during several periods. As the game goes on, brigand bands (run by the computer) will become an increasing threat to your fiscal (and physical) well-being.

Taxes were raised, somewhat inefficiently and often with such vigor that rebellion resulted. The brigand depredations could also get the common folk riled up. It's bad enough when they refuse to pay their taxes, but if you're a real bad actor, they'll come after your head.

If you decide to spend all your time fighting in wars and tournaments, your fiefs will "run themselves," although less efficiently. In other words, without your personal attention, your bailiffs will be less rigorous and honest. Your wife and children may develop excessive spending habits and, in general, your assets will waste away.

You can borrow and lend money, or buy and sell fiefs. Fiefs are the basic economic unit of the game, and there are about 800 of them. The "fief file" is itself an interesting historical document, containing data dredged up from numerous sources and containing traditional "fiefs" (agricultural areas) as well as towns and religious facilities (church-owned fiefs with monasteries, etc.). There is, for example, a fief in France called Condom, run by a player who will be known as the Sire de Condom. There's also another Welsh fief called something worse, but it's in Gaelic. Hey, we just call them as we see them. But you can see where the French got some of their historical traits from.

Play the game, and you'll quickly discover that even the largest cities had fewer than 100,000 citizens within their walls. Less than 5 percent of the population lived in walled towns and cities; most of the remainder lived in small villages surrounded by farmland and pastures.

About 2 percent of the population belonged to the nobility or a noble's household. This group included many of the troops. Thus, of the 30 million (pre-plague) population in the area covered by the game, only a few hundred thousand men were fighters. The 300 players occupy one of three ranks. Every player is at least a "lesser noble" controlling one or more fiefs. Three dozen players are "magnates," or overlords of several lesser nobles and themselves the holders of more fiefs than a lesser noble. One player is the king of England, and one really unfortunate wretch is the king of France.

To start the game off in the proper spirit, the English players (about one-third the total) are allowed a week to discuss things among themselves and elect those of their number deemed most suitable to be magnates and the king. This represents the superior cohesiveness of the English government, even though the English did not have elections as such. Those players who chose to be French (or were forced to be French if all the English slots were taken) are randomly assigned as lesser nobles, magnates, and king. This maximizes the chances of the wrong person getting into a key position.

While combat is the final arbiter of events, there are other ways to get things done. After all, battles are only a forceful way of obtaining a treaty. The English king wants to become the French king, and to do so, he must control enough French territory to make his coronation as French king convincing. There is a legal system built into the game, covering both civil and church matters. You can be declared an outlaw or excommunicated. You can even be tried and executed. There's a pope (actually, more than one during this period), heresy, and the ever-popular Black Death (which will ultimately destroy half the population over several outbreaks).

Money plays a large role in the game. Although several currencies were used in this period, we invented a new one, the ducat (which was actually the name of an Italian currency) so that you wouldn't have to deal with all the different exchange rates. By way of example, 600 ducats equals one English pound and 135 ducats equals one French livre, and so on. One ducat also equals one U.S. dollar (1992 vintage). Money is important in this period. Troops were the biggest expense. The feudal levy could still be called out, but only for local defense and only for a limited period of time. On average, a mercenary soldier cost about 1,000 ducats per turn (a season of three months). Most armies in this period contained 5,000–10,000 men, which meant 10–20 million ducats per turn just for the troops. The king of England had only that much income per turn (in the best of times), and there were other expenses (household, maintenance of fortifications, bribes and payments to officials, etc.). The king had several thousand people on his payroll, and even a magnate had several hundred servants and soldiers to support (and an income of up to several million ducats per turn). You can borrow money, but basically the English kept the war going for so long because they were better soldiers, their ships controlled the English Channel (most of the time), and they were able to constantly plunder French fiefs. Thus, the object of taking your army into the field was not battle, but plunder. Sort of like a game of Monopoly with edged weapons.

Using a remote computer does not waste the capabilities of your PC. For players with IBM-compatible machines and Macintoshes, there is a program you can download from GEnie. This program (a Graphic Front End or GFE) will incorporate a datacomm program to connect with GEnie (more specifically, the *Hundred Years War* game) and have an editor for writing messages to other players and the graphics capability to display game activities in more detail. Players with other PC types can still play just by connecting with GEnie, but will not have access to all the amenities in the GFE. The GFE costs nothing (except the connect time to download it, a few bucks).

When you first connect with this game via GEnie, you immediately know you're not in Kansas anymore. If it's your first time, you're asked if you want to join the game. If you answer yes, you receive a screen full of information on who your character is ("You are William de Clinton, the Baron of Huntington in Cambridgeshire, England, etc.," or "You are the Captal de Buch, lord of the fortifications guarding the port of Buch on the Gascon coast and the owner of three fiefs, etc.") and what your resources are (other fiefs owned, annual income, etc.), as well as personal information (age, state of health, wife, kids, state of your fortifications, troops under your control, claims on other players' fiefs, etc.).

You can sign on and do nothing, and your character's life will go on. In this case, you would be the medieval equivalent of a couch potato. Taxes are collected and spent in your estate, dishonest officials will embezzle most of the surplus, and life will go on with you doing little more than observing. Note that you play through the current eldest male of the family your character belongs to. When the eldest male dies, the next eldest male descendant takes over. If none is available, the line dies, and you're out of the game. Among other things, you have to look after the wife and kiddies (which involves some interesting and obvious activities not normally found in wargames but essential to medieval military affairs).

However, there is much to do, and this is the virtue of an on-line computer game. The key here is the multiple players and the ease of communication and interaction. Multi-player games are usually more enjoyable than two-player encounters, yet solitaire play is the most common because of the difficulty in getting the players together. On-line games eliminate these problems. Whenever you get on, you can send and receive messages. You can also talk with other players who are on at the same time and join together for tournaments (jousting, duels, etc.), hunting or conspiracies or affairs of state (there was often little difference between the two). This form of real-time communication, in a time when it took three weeks to ride from one end of France to another, is justified by the three-month length of the game "turns" (one day of real time). Players may allocate their 90 days more productively, performing such drudgery as managing their fiefs or training with arms. But true to the period, they may spend most (or all) of their time out hunting and fighting. Naturally, the more ambitious (and probably the older) players will more readily adapt to building their power through management and diplomacy. The important thing is that up to 300 individuals in one historically based game will make for an interesting experience. Should some bloody-minded sixteen-year-old find himself or herself king of France, surrounded by a few equally rapacious nobles, and faced with a thirty-five-year-old tax accountant, management consultant, or college professor playing the king of England—

all the better. Well, this game will be fun just to watch. And you will be able to watch. For example, we can have play-by-play descriptions.

Battles during the period were infrequent (perhaps one or two a month of real time, or one or two every seven years of game time). However, you could usually see them coming (money and mercenaries had to be collected, negotiations conducted, etc.), so we will announce them and enable any interested GEnie users to view the action in terminal mode. "*Hundred Years War:* Witness the Battle of Rouen at 9 PM, February 3rd, 1993." The "Heralds" (sysops) will give play-by-play commentary, battle maps (*O*'s and *X*'s) plus participant interviews, pre- and post-battle commentary, and the like. Everything but tailgate parties. Sort of like a football game with edged weapons, no time-outs, and no referees (and very few rules).

"Here we are, folks, outside the French city of Rouen in the year of our Lord 1376, where the duke of Normandy and the count of Artois have marshaled their forces to settle a dispute over who shall be the duke of Brittany. The Normans have a contingent of English troops under the earl of Bedford and are the current favorites. The six thousand Norman troops and eleven thousand French troops are arrayed, and combat is expected to commence in about fifteen minutes at nine-forty P.M. EST. First, however, we have a live interview with his lordship, the constable of France, the count of Artois. Tell us, Count, how do you expect to overcome those English longbowmen?" And so on . . .

More frequently, there will be tournaments. These involve all sorts of individual combat, including jousting. The joust will be handled like the familiar vehicle simulators computer games are famous for. Our version will have dozens of players from all over the country lining up electronically to come thundering down the lists in real time for fame, glory, and (game) money. We're also working on a system whereby other players can watch the action, and perhaps even make bets (with ducats only) on the action (another bit of historical realism). Note that jousts were also a popular way to settle sieges. Real (not blunted) weapons were used, and the fight was to the death.

Naturally, new technology costs more. Connect time for on-line games via the GEnie network is six dollars an hour (as of 1992, billed to the nearest minute). The game is being designed so that, if you use the GFE, you can actively participate using only two to three hours a month. If the game goes a full century (400 days), it's going to cost you something like $200 to play the entire game. That's 50 cents a day. Not cheap, but not all that outrageous either. How else could you participate in the Hundred Years' War? One of the big problems with on-line games is their expense. Many players find themselves spending hundreds of dollars a month on these addictive games, and many have to drop out because they can't afford it. *Hundred Years War* will be unique in that it will be one of the first on-line games to confront this problem. The way the game is designed, there is no real advantage to spending hundreds of dollars a month on the game (unless you want to be really good at jousting, which is actually a relatively minor aspect of the period). Players can have a major impact while only incurring costs of no more than $20–$30 a month. This way, we hope to get a lot more players involved. On-line gaming looks to be one of the more interesting new areas for historical simulations in quite some time.

But why just talk about it when we can walk you through a few activities of the actual game (alpha version) in late 1991? I was playing Heinrich V, Graf de Limbourg. What follows is what actually appeared on my computer screen as I connected to GEnie and entered the *Hundred Years War* game. Brief notes [in brackets] explaining what I'm doing will appear interspersed with the game related material.

---

**** Thank you for choosing GEnie ****

**The Consumer Information Service from General Electric**
**Copyright © 1992**

**GEnie Logon at: 05:46 EST on: 921107**
**Last Access at: 21:47 EST on: 921107**

**No letters waiting.**

**GEnie Announcements (FREE)**

1. NEW FCC COST INCREASE THREATENS COMPUTER SERVICES ................ *FCC
2. NEW NEW NEW The 1992 Version GEnie Satin Jacket ..................... *ORDER
3. New Area Codes in California and Maryland ........................................
4. If you love Soap Operas, check out the >free< RTCin .................. SHOWBIZ
5. Subscribe now to DUNGEONS & DRAGONS(tm) games .......................... TSR
6. November issue the Caribbean Travel Roundup is here .......................... TIS
7. Win MICROSOFT SOLUTION software at FREE RTC .......................... HOSB
8. RTCs, Mickey's Birthday ............................................... FLORIDA
9. RECHARGER Magazine RTC on the ........................................ PSRT
10. Computer Football League opens ..................................... SCORPIA
11. Speak with a Technical Support Supervisor .......................... CHIPSOFT
12. LIVE from LEIPZIG ................................................... GERMANY
13. Mid-Week football fun. Call the plays for Texas A&M .......................... QB1
14. Play the weekly Fantasy Football Pool and win $$ ................................ FF
15. Upload contest wins FREE time, Usenet news in .............................. UNIX

Enter #, <H>elp, or <CR> to continue? Move 945

[The first thing you see is GEnie ID and promotional material. You are not charged for looking at that. I entered the command "Move 945," and that took me to the *Hundred Years War* (*HYW*) game. The GFE would take you there automatically.]

The game and its RT are a GEnie VALUE ($6 an hour) Service

WELCOME TO THE HUNDRED YEARS WAR MULTIPLAYER ONLINE GAME

The game is on page 945, the HYW RT is on page 946.
Weekly Real Time Conferences on Saturday, 4 PM Eastern Time, in the games Interactive Court. Plenty of files on the game in the file library (page 946). Plenty of good advice and companionship in the HYW RT (page 946). Plenty of mayhem and adventure in the game (page 945).

HYW Staff, and their GEnie E-mail IDs

Designed by Jim Dunnigan: HYW$
System development by Dan Masterson: HYW$
Research by Al Nofi: HYW$
Player assistance provided by the Heralds:
Herald of England- Darrell L. Killpack: FROTZ
Herald of England- Mark O. Kinkead: M.KINKEAD
Herald of France- Barbara Byro: BYRO
Herald of France- Robert B. Kasten: R.KASTEN1
Herald of the Others- Daniel H. Sceltema: D.SCELTEMA
Master of the Archives- Charles R. Townsley: C.TOWNSLEY
Master of the Peerage- Richard A. Edwards: R.EDWARDS26

Help is always available in the HYW RT (page 946)

GEnie
Page 945

Hundred Years War (tm) by DENO

1. **Personal Affairs**
2. **Travel**
3. **Hunting, Combat and Dirty Deeds**
4. **Official Acts**
5. **The Herald**
6. **Scoreboard**
7. **Instructions**
8. **Join Hundred Years War**
9. **HYW RoundTable**
10. **Send FEEDBACK to DENO**

Enter #, <P>revious, or <H>elp?1

[This is the main HYW menu. The first thing I want to do is go to the personal affairs menu (1) to see how I'm doing.]

Personal Affairs Menu

1. **<Fie>f Management**
2. **<P>ersonal Characteristics**
3. **<H>ousehold Affairs**
4. **<Fa>mily Matters**
5. **<Fin>ancial Activity**
6. **<C>ourt**
7. **<A>larum Menu**
8. **<G>amble**
9. **<O>nline Messages**
10. **Dump Vital Financial Information**
11. **Dump Vital Family Information**
12. **<I>nteractive Court**

Enter #, or <ENTER> To Exit: 2

[Personal affairs has numerous submenus. But first I'll show you "who I am" by choosing menu option 2.]

Fall Of 1340 90.0 Days Left

Personal Purse 877.00
Current Location is AFC01 Luxeil
Current Health 3 Maximum Health 3

| | | | | | | | |
|---|---|---|---|---|---|---|---|
| Management | 3 | Guile | 1 | Leadership | 8 | Stature | 7 |
| Protection | 2 | Endurance | 8 | Attack Value | 1 | | |
| Tournament | 3 | Sire ID | 0 | Org ID | 285 | | |

You are Heinrich V de Limbourg     Your ID is 285     Your age is 46
You are married to Alais de Limbourg.
Sex m   You speak F1
Wife is not Pregnant

Court Persona Is 285

| | |
|---|---|
| Athlete | 4 |
| Command | 7 |
| Insightful of people | 8 |
| Insightful of situations | 4 |
| Superstitious | 5 |

See Titles (Y or N)? y

| Fief ID Fief Name | Rank |
|---|---|
| 1. HLG01 Limbourg | 11 Graf |

Press <ENTER> to continue.

[If it looks like something out of a role-playing game (RPG), it is, as *HYW* is an RPG, albeit one played on a vast scale and historically accurate. I describe my character as a dirty old man, but more on that later.]

[Back at the personal-affairs menu, I go to check out my financial situation. As in real life, money drives this game.]

Financial Activity

1. **<S>ummary of all Holdings**
2. **<I>ndividual Fief Status**
3. **<F>iefs Owned List**
4. **<C>urrent Finances**
5. **<P>ayments to Other Players**
6. **<B>uy and Sell Fiefs**
7. **Hire and Fire <N>PCs**
8. **Hire <T>roops**
9. **<E>xit**

Enter # 3

[Let's check out my fiefs, the source of a feudal lord's wealth.]

| | Fief ID | Fief Name | Trsry | Kp Lv1 | Surp | Llty | Bail | Mngr | AT | Status |
|---|---|---|---|---|---|---|---|---|---|---|
| 1. | HLG01 | Limbourg | 6 | 5.79 | -35 | 9.0 | 392 | 0 | N | Calm |

Choose Fief 1 to 1, <E> Followed by Fief # to Examine, <P>revious Page,
or <ENTER> TO Quit: 1

Fall Of 1340 90.0 Days Left

Current Fief HLG01 Limbourg
Loyalty 9.00 Surplus -35.00 Treasury Balance 6.23 Status Calm
Bailiff ID 392
Auto Transfer Flag Is Off, Surplus Will Not Be Transferred To Purse

Personal Purse 877.00

[The Graf de Limbourg controls one fief, with a name and location, that still exists in modern Germany. This was one of the many peripheral areas included in the game. Most of the fiefs are in France and England.]

Enter # 1

Individual Fief Summary

Limbourg (HLG01 ) Liege HRE:Germany Population 20.3
Language F1 Freedom 1 Status Calm
Your Overlord is Guy Baudet (290) Bailiff ID 392
Fields 5.22 Industry 2.08 Weather 0.99 Trsy Bal 6.23
Knights 28 MAA 10 Lt Cav 0 Yeomen 0 Foot 406 Rabble 8932

| | Last Season | Crnt Season | Next Season |
|---|---|---|---|
| Loyalty | 8.73 | 9.00 | |
| GDP | 3117 | 3121 | 3138 |
| 1-Tax Rate | 14.0% | 14.0% | 14.0% |
| Income | 436 | 436 | 439 |
| 2-Officials | 80 | 66 | 60 |
| 3-Garrison | 222 | 182 | 165 |
| 4-Infrastructure | 122 | 100 | 90 |
| 5-Keep (Level) | 160 (5.64) | 131 (5.79) | 119 |
| Extra Expenses | -24 | -7 | |
| Total Expenses | 560 | 471 | |
| Graft | 0 | 0 | |
| Overlord Taxes | 0 | 0 | |
| Surplus/Deficit | -124 | -35 | |

Enter 1–5 to change, or <ENTER> to Quit:

[I'm currently taxing this fief at a high level, but I'm also investing a lot in infrastructure (roads, public buildings, and the like) and payments to the garrison (local knights and men at arms), which will enable me to raise taxes even higher.]

[Back at the main menu, I go to the alarum menu, to find out what has been happening in the game recently. Each day of real time represents 90 days (one season) of game time.]

Alarum Menu

1. **<R>ead Personal History File**
2. **Read <H>erald**
3. **<A>ctive Player List**
4. **<S>cores**

999. Move to Main HYW Menu

Enter #, or <ENTER> To Exit: h

Enter Season Of History You Want To Read

1. **Last Season .................................................. (Summer 1340)**
2. **Two Seasons Ago ............................................ (Spring 1340)**
3. **Three Seasons Ago ........................................ (Winter 1339)**

Enter #, or <ENTER> To Exit: 1

[I look at the most recent season (yesterday in real time) and find out who was born, who died, who is at war, what battles were fought, keeps besieged, and fiefs pillaged. Never a dull moment in the 14th century.]

History For Summer 1340

The Freville Family (18) has had a baby boy.
The Burys Family (46) has had a baby boy.
The de Bohun Family (101) has had a baby boy.
The de Camus Family (119) has had a baby girl.
778 Catherine Berkeley died in child birth.
The Glyn Dwr Family (195) has had a baby girl.
658 Marie de Preaux died in child birth.
The de Rohan Family (205) has had a baby girl.
The du Barril Family (206) has had a baby boy.
The Holland Family (226) has had a baby girl.
The Douglas Family (282) has had a baby boy.
The de Loraille Family (297) has had a baby girl.
The de Namur Family (305) has had a baby girl.
The Grovesner Family (239) has had a grandson.

[Because you only can stay in the game if you have an heir to replace your current character when it dies, marriage and children are important. Childbirth was, however, more dangerous then than it is today and even the wives of aristocrats were at risk.]

Eudes de Burgundy (53) has died at Dijon.
Isabelle Richemont (103) has died at Limoges.
John de Cobham (163) has died at Maidstone.
Simone Boccanera (287) has died at Genoa.
Andrea Orsini (407) has died at Venice.
Louise de Craon (687) has died at Rochefort.
Jeanne de Laval (705) has died at La Fert.
Margaret ap Gwain (715) has died at Montbard.
Press <ENTER> to continue.
Jeanne de Clare (716) has died at Llandovery.
Annette de Gonzolles (760) has died at Llanbyther.
Thomas de Floques (950) has died at Portsmouth.
Alexandre de Corvino (1235) has died at Evreaux.
Eudes du Barril (6274) has died at Perigord.
Alain de Lyon (6315) has died at Laon.

[Warfare was common, and often the forces were commanded, if only in name, by women. These female warriors were almost always widows who were forced to send out troops to defend their interests. The women would usually hire a noted free-lance commander to actually lead the troops in battle. Unlike most wargames, this one actually attracts a large number of female players.]

FP003 La Roche de Poitiers is being besieged by Anne Aubert (112).
FP003 La Roche de Poitiers was taken by Anne Aubert (112).
APR01 Barcelonette is being besieged by Jean de Clermont (121).
APR01 Barcelonette was pillaged by Jean de Clermont (121).
APR01 Barcelonette was raided by Jean de Clermont (121).

APR01 Barcelonette was taken by Jean de Clermont (121).
ANC02 Puget is being besieged by Jean de Clermont (121).
Mary Elizabeth Clifford (230) and Robert de Nesles (764) were married.
Anne Glyn Dwr (6303) and Roger Mowbry (1079) were married.
Jean de Clermont (121) Has Been Excommunicated.
Blanche de Ponthieve (632) and Charles de Chabannes (1127) were married.
Guy de Sully (82) Has Issued A Call To Arms.
Bodo Badarieux (1583) was caught attempting to assassinate Gautier le Roy (60).
Herve Gex (3689) was caught attempting to kidnap Jeanne Le Roy (799).

[Assassination and kidnapping were considered perfectly reasonable ways to achieve your goals in this period. They were a lot cheaper than hiring an army.]

APR01 Barcelonette is being besieged by Clare Paleologo (231).
Press <ENTER> to continue.
Jean de Clermont (121) Has Been Outlawed
By Philippe VI de Valois (200).

[At this point in the game, the player with the Jean de Clermont decided to wage a private war. The French king (Philippe VI de Valois) told him to stop, as did the player playing the Pope. De Clermont ignored both, and in this season he was outlawed by the king and excommunicated by the Pope. The king and his loyal vassals raised armies and marched on de Clermont's lands in the south central French Forez region. Doesn't pay to mess with the king.]

FCE05 Mantes was pillaged by Ame de St-Vollier (10)
FCE05 Mantes is being besieged by Ame de St-Vollier (10).
FCE05 Mantes was raided by Ame de St-Vollier (10)
Eleanor de Grailly (1043) and John d'Urtino (531) were married.
Roger de Clermont (789) was captured during a successful siege.
Elizabeth de Clermont (6228) was captured during a successful siege.
FCE05 Mantes was taken by Ame de St-Vollier (10).
FCE03 Pontoise is being besieged by Ame de St-Vollier (10).
FCE01 Clermont was raided by Thierry III de Grand Pre (252)
FCE01 Clermont is being besieged by Philippe VI de Valois (200).
Jean de Clermont (121)'s Garrison Defeated
Thierry III de Grand Pre (252) During A Pillage/Raid Attempt At Clermont.
FCE01 Clermont surrendered to Philippe VI de Valois (200)
FCE03 Pontoise was pillaged by Philippe VI de Valois (200)
APR01 Barcelonette is being besieged by Louise de Gonzolles (173).
FCE03 Pontoise is being besieged by Philippe VI de Valois (200).
FCE01 Clermont is being besieged by Thierry III de Grand Pre (252).
Thierry III de Grand Pre (252) Defeated
Philippe VI de Valois (200)'s Garrison During a Siege Attempt At Clermont.
FGU12 Graves is being besieged by Jean de Grailly (126).
Press <ENTER> to continue.
Jean de Grailly (126) Defeated
Foucaud de Rouchechourt (3)'s Garrison During A Siege Attempt At Graves.
Jean de Grailly (126) Defeated
Foucaud de Rouchechourt (3)'s Garrison During A Siege Attempt at Graves.
Foucaud de Rouchechourt (3) was captured during a successful siege.

Bodo Digne (3257) was captured during a successful siege.
FGU12 Graves was taken by Jean de Grailly (126).
FCE01 Clermont paid extortion to Thierry III de Grand Pre (252)
Thierry III de Grand Pre (252) Defeated
Philippe VI de Valois (200)'s Garrison During A Siege Attempt At Clermont.
Thierry III de Grand Pre (252) Defeated
Philippe VI de Valois (200)'s Garrison During A Siege Attempt At Clermont.
Anne de Breche (1120) and Nicholas de Breche (186) were married.
APR01 Barcelonette was pillaged by Alfonso XI de Castilla y Leon (8)
APR01 Barcelonette was raided by Alfonso XI de Castilla y Leon (8)
APR01 Barcelonette is being besieged by Alfonso XI de Castilla y Leon (8).
APR01 Barcelonette was taken by Alfonso XI de Castilla y Leon (8).
Elizabeth de Bertrand was caught seducing
Bertrand du Guesclin.
Elizabeth de Bertrand (193) and Bodo de Mauny (794) were married.
Henry Percy was caught seducing
Annette de Stafford.
Press <ENTER> to continue.

[Back at the alarum menu, I can also see who the other players are. In this game there were over a hundred at this time. Below is one screen full.]

Active Players

| | ID | Name | | E-Mail |
|---|---|---|---|---|
| 1. | 22 | **Benedict XII** | **Pontifex Maximus** | **GM** |
| 2. | 83 | **Guy** | **Baveux** | **SIMUTRONICS** |
| 3. | 222 | **Renard VI** | **de Pons** | **FDITIZIO** |
| 4. | 109 | **Guy** | **d'Albon** | **J.JIMENEZ** |
| 5. | 58 | **Gaston II** | **de Carcassone** | **CGW** |
| 6. | 47 | **Edward III** | **Plantagenet** | **DIPLOMACY-1** |
| 7. | 60 | **Gautier** | **le Roy** | **DIPLOMACY-3** |
| 8. | 176 | **Louis** | **de Bourbon-LaMarche** | **AUSI-SUPPORT** |
| 9. | 364 | **Connor** | **McKinnon** | **A** |
| 10. | 97 | **Hugh** | **de Audley** | **FROTZ** |
| 11. | 288 | **Personne I** | **Inconnu** | **M.WIELENGA2** |
| 12. | 46 | **Edward** | **Burys** | **J.BRANDT7** |
| 13. | 157 | **Juana II** | **de Navarra** | **B.HUNTER7** |
| 14. | 12 | **Anger** | **de Montault** | **J.CUMMINS** |
| 15. | 164 | **John** | **Mowbry** | **W.HART9** |

Choose Character 1 to 15, <N>ext Page, or <ENTER> TO Quit:

[Back at the main menu, I go to the travel menu.]

Travel Menu

Fall of 1340 86.0 Days Left

----------7-Northwest----------
- AFC02 Vesoul 1.0-

------------------9-Northeast-----------------
- AFC01 Luxeil 1.0-

-------------4-West-------------
AFC05 Pesmes 1.0-

Current Fief
AFC04 Besancon

--------------6-East-------------
AFC03 Clerval 1.0-

-------1-Southwest--------
- AFC07 Salins 1.0-

----------3-Southeast---------
- AFC06 Pontarlie 1.0-

You Are Outside The Keep No Army Present In Fief

Choose Number Of Direction You Wish To Move, or

2. <Exa>mine Fief
5. <V>isit Court and Enter Keep
8. <Po>rt Movement
10. <A>rmies in Fief
11. <Ent>er Keep
12. <Exi>t Keep
13. <L>ist Those Outside Keep
14. <C>ombat\Dirty Deeds Menu
15. <F>ief Management
16. <Pe>rsonal Affairs Menu
17. Army <M>anagement
18. <O>nline Messages

Enter #, or <ENTER>To Exit: v

[The travel menu is an easy way to move around Europe. Note the use of the six-sided ''hexagon'' technique to regulate movement. Just like many manual games and many computer wargames also. The travel menu also allows access to nearly all the other features of the game. I decide to visit the court of the fief I am already in.]

[I choose to ''attend court'' to see who is hanging around the local lords' château. Most of the characters here are NPCs (nonplayer characters).]

Court

Fall of 1340 86.0 Days Left

Fief AFC04 Besancon
The Owner Is Jeanne II de Bourgogne (156) Who Is Active.
Province Franche-Comte Kingdom HRE:Arles
The Overlord is Jeanne II de Bourgogne (156)

| | ID | Name | | Org | Sex | Type | Comp |
|---|---|---|---|---|---|---|---|
| 1. | 156 | Jeane II | de Bourgogne | 156 | F | PLYR* | |
| 2. | 465 | Thomas | de Beauchamp | 156 | M | FMLY | |
| 3. | 475 | Alais | de Limbourg | 285 | F | FMLY | YES |
| 4. | 499 | Philippe | de Bourgogne | 156 | M | FMLY | |
| 5. | 915 | Robert | de Savoy | 156 | M | FMLY | |
| 6. | 939 | Clare | de Savoy | 156 | F | FMLY | |
| 7. | 1015 | Ame | de Savoy | 156 | F | FMLY | |
| 8. | 1611 | Roger | Baiona | 156 | M | NPC | |
| 9. | 2038 | Bernard | Bourg | 156 | M | NPC | |
| 10. | 2719 | Sean | Chateau-Renault | 156 | M | NPC | |
| 11. | 2753 | Bernard | Chaumont | 156 | M | NPC | |

Choose Character 1 to 11, <E> Followed By # To Examine, <N>ext Page,
or <Q>uit: e6

[Well, well, it seems I have come upon the family of the owner. There is Jeanne II de Bourgogne, obviously the widow who now runs the place. The others belonging to "Or-g(anization)" 156 are her children, and Thomas de Beauchamp must be the husband of one of her daughters. I think I'll get to know one of the daughters.]

This is Clare de Savoy (939) who is Married.
She is Not Pregnant
Current Health 5 Maximum Health 5
Age 21 Sex f Language F2 Loyalty 3

| | | | | | | | |
|---|---|---|---|---|---|---|---|
| Management | 8 | Guile | 4 | Leadership | 1 | Stature | 1.0 |
| Protection | 1 | Endurance | 3 | Attack Value | 2 | | |
| Tournament | 3 | Sire ID | 177 | Org Id | 156 | | |
| Berserker | | | 1 | | | | |
| Collector | | | 4 | | | | |
| Evil Eye | | | 7 | | | | |
| Insightful of people | | | 9 | | | | |
| Insightful of Self | | | 2 | | | | |
| Keepmaster | | | 1 | | | | |
| Siegecraft | | | 6 | | | | |
| Sorcery | | | 6 | | | | |

Press <ENTER> to continue.

[Hmmm, young Clare is certainly a piece of work, great manager, lousy leader, and a taste for the occult. A married witch, let's see if we can get acquainted. . . .]

Current Character: 939 Clare de Savoy

1. **<E>xamine Character**
2. **<H>ire NPC**
3. **<S>educe**
4. **<A>dd To Traveling Companions**
5. **<R>emove As Traveling Companion**
6. **<ENTER> To Quit**

Enter # 3

Sire, what type of lady do you take me for!
Press <ENTER> to continue.
Seduction Takes One Day.
Year: 1340 Season: Fall Days Left: 85.0
Press <ENTER> to continue.

[Hmmm, these young wives can be unpredictable. But let's try again.]

Enter # 3
Let us steal away to a place of peace and quiet
Press <ENTER> to continue.
Seduction Takes One Day.

Year: 1340 Season: Fall Days Left: 81.0
Press <ENTER> to continue.

[Ahhh, much better. Chivalry lives in the shadows. Hanky-panky was included as a game function because it was quite common in the 14th century. In fact, you can't have a realistic game of the Hundred Years' War without adultery. It seems that a major problem the French had was the ruling family, the Valois. At this point in time the Valois were "genetically challenged." As the game simulates the passing of characteristics by both parents to their children, it would have taken several generations of outstanding Valois wives to breed all the bad characteristics out of the line. France could not wait that long. One of the French kings halfway through the war was totally loopy and his wife was receptive to the attentions of other nobles. This resulted in a Valois heir that was not a Valois (the queen publicly admitted as much). France noted that the illegitimate Valois was a much smoother article than the genuine Valois and accepted the bastard heir. This did much to turn things around for the French.]

[Back at the travel menu, I go to the army management menu; I proceed to hire some troops by recruiting in the current fief. This I can do immediately by offering local troops money to follow my banner. This cannot gather many troops, but when you need some soldiers in a hurry, this is the way to go. Gathering a large army—the "Call to Arms"—takes several months and everyone knows about it.]

Army Management

1. **<C>all To Arms**
2. **<R>ecruit In Current Fief**
3. **<A>rmy Status**
4. **<D>isband Army**
5. **<M>uster Out**
6. **<S>tanding Order**
7. **<T>ransfer Troops**
8. **<P>ick Up Troops Responding To Call To Arms**
9. **Assemble <F>leet**

999. Move to Main HYW Menu

Enter #, or <ENTER> To Exit: r

Purse: 877.00 Kducats

Soldiers cost 1 Kducat Per Soldier Per Season.
You must, also, pay a 1 Kducat recruiting fee.
How many soldiers do you wish to hire: 200
You recruited:
16 Knights
7 Men At Arms
0 Yeomen
177 Foot
Your recruiting took 3 days.

Year: 1340 Season: Fall Days Left: 76.0

Press <ENTER> to continue.

[I've got my troops, who will stay with me as long as I pay them.]

[There's a lot more to the game. But you have to be there to appreciate it all.]

[Back at the main menu, I leave the game and GEnie.]

Thank you for choosing GEnie.
Have a nice day!

Online: 9 minutes, 50 seconds.

OFF AT 21:56EST 11/07/92

NO CARRIER

## COMPUTER WARGAME DESIGN TIPS FOR THE MILITARY DESIGNER ✷✷

There are several items that the military-wargame designer has to consider. These are issues that the commercial-wargame designer rarely, or never, encounters. To put it in the proper perspective, I call wargames "models" in this section. Especially in the military, "game" is a dirty word.

### MAP OR NOT

A key decision is whether your model needs a map display. Some don't, but nearly all can benefit from one. If not a tactical map display, then graphic displays to make data easier to absorb.

### VISIBLE OR BLACK BOX

One easy way to cut costs and trim development time is to minimalize user access to what is going on in the model. Models with limited access are known as black-box models. In the past, when computer resources were limited, or unavailable, black boxes were unavoidable. The downside was that black boxes are less credible to users, particularly if these results are striking and against the grain. The visible model allows the user ready access to what's going on in the model. The user can then, often interactively, query the model as to how a particular result was reached.

### SANITIZED BASIC VERSION

Most defense-community models are highly classified. A major shortcoming of this is the limited number of people who can review the model. The greater the number of reviewers, the more likely it is that errors and problems in the model will be detected and corrected. This problem can be corrected by developing a version of the model that is unclassified (sanitized).

### INTELLIGENT USER ACCESS

It's not enough to give the user access to the inner workings of the model; it must be efficient, thorough—in other words, intelligent access. Makes both the model and the user much more efficient and effective.

### INEXPENSIVE WORKSTATIONS

What was an engineering workstation five years ago is today a high-end, 32-bit personal computer costing less than $5,000 a unit with hard disk and high-resolution graphics. This provides the opportunity to create a model that can be used very widely. This is an asset that must be considered.

# WHO PLAYS THE GAMES

Wargaming is the hobby of the overeducated. Nearly all the adult gamers are college graduates, and most of those also have advanced degrees. They are also eager and forthcoming when asked about themselves.

While at SPI, I developed a market-research system that constantly monitored gamer demographics, new-product preferences, and existing-product satisfaction. The questionnaires were (and still are) a regular part of *Strategy and Tactics* magazine, as well as several other wargaming periodicals. It became something of a tradition. I have access to much of the survey results for the 20-plus years *S&T* has been running these surveys. Through this period, due to these regular surveys of wargamers, a large number of interesting bits of information have been collected. Many of the questions asked were generated by little more than a sense of curiosity, but some of the questions and their answers should prove interesting to gamers as a means of defining who they are as a group. The answers not only tell us who current wargamers are, but how they got that way.

These trends were uncovered by using a correlation and factor analysis. Only statistically significant trends are commented upon here. In addition to periodic random surveys to validate the voluntary results, we also performed correlation analysis on new-game preferences (before the decision to publish was made) and subsequent sales. It was a nearly perfect fit. Even I was impressed. Pity I wasn't publishing mass-market items. But then, we probably wouldn't have got such thoughtful feedback in the first place.

Taken as a group, wargamers were relatively young 10 years ago, when 52 percent were under age 22. Today, only a few percent are that young, and more people are getting into wargames at an older age, typically in their 20s. Wargamers are predominantly male, although the percentage of women players increased when computer wargames appeared. While paper wargamers are about 1 percent female, computer wargamers are 2 to 3 percent female.

When gaming was a relatively new phenomenon, most players were quite young. The fact that most of them were in school, where other students could easily find out about wargaming, certainly helped the spread of the hobby. When role-playing games (RPGs) became available, the social networking of students worked against wargamers. The kids who played wargames were generally the brightest, if not always at the head of their class. RPGs are easier to get into, and a much larger number of students were able to participate. Many of the wargamers became gamemasters, the one participant in an RPG who has to keep track of a lot of things simultaneously. Wargamers have a lot of skill and experience at that. Moreover, the average age of getting into wargames was 12 or 13 years old. Wargames are, intellectually, an adult exercise, and younger kids really couldn't hack it. RPGs, however, can be used and enjoyed by children several years younger than that. It's no wonder a lot of guys in their 20s, having seen nothing but RPGs for the past 10 or 15 years, were thrilled to eventually discover wargames that had been overshadowed by the greater popularity of RPGs. In many ways, the enthusiasm for wargames is similar to an enthusiasm for books. It tends to stay with you.

During the 1970s, when there were a lot of youngsters coming into wargaming, there were also a lot of them leaving it when they got older. College, marriage, a job, all tended to create an average of six years spent wargaming. The positive side of a lot of teenagers not coming into gaming was that those who did get into wargaming tended to stay longer, with the average of duration nearly tripling. Today, the majority of wargamers have been at it for over 10 years. Another odd aspect of the age distribution of wargamers is the lack of people born before World War II. Now, one can understand a lower range limit of entering wargamers. But consistently, from the late 1960s to the early 1990s, very few potential wargamers who were born before December 1941 become wargamers. There's a real generation-gap situation there. It doesn't make much sense, but there it is.

The education profile follows the age profile. In 1980, 13 percent of wargamers were high school students, 9 percent high school graduates, 13 percent had attended college, 12 percent were college students, 16 percent were four-year-degree graduates, and 37 percent had either graduated from graduate school or were still attending it. When one separated the school-age population from the older folks, one found that more than 50 percent of the older gamers possessed 17 or more years of education. This pattern held up through the 1980s and into the 1990s. In 1980, 10 percent of wargamers were students; most of the remainder had graduate school or better education. In the early 1990s, less than 5 percent of wargamers are students, and more than half the remainder have more than 16 years of education.

Even considering their education levels, gamers read a lot. Over half belong to book clubs and an even greater proportion belong to, as one would expect, a military book club. A smaller percentage belong to a science-fiction book club. This puts them way above the average in book-buying, and reading, habits. Statistically, gamers also score high on magazine readership: nearly 50 percent read *Scientific American*, about a third read science-fiction magazines, and over half read computer-related magazines. Nearly 30 percent subscribe to a news magazine, and nearly as many read comic books. Considering some of the exquisitely detailed flak I've got over data errors in games, I've sometimes wished wargamers spent more time watching football. The fact that the first edition of this book stayed in print for over 10 years, and led to a second edition, testifies to the book-buying prowess of the relatively small universe of wargamers.

Older and more educated gamers prefer less complex games and have fewer hours to spend playing the games. But at the same time, more experienced gamers have, well, more experience with complex game systems and are thus comfortable with complex games. As I always point out to aspiring game designers, the key to designing a game is thorough knowledge of all the design techniques used in the past. The most common way of getting that knowledge is playing and studying a lot of different games. Someone who has been playing games for 10 or more years has a different concept of what is simple than someone who is new to wargames. Wargames have always been arcane, but now publishers are putting out "simple" games that are regarded as such only by the more experienced players. If a game that is simple in absolute terms is published, the experienced gamers who comprise the majority of the buyers turn their nose up at it. This has caused problems in providing simple enough games for newcomers to hone their wargaming skills on.

The more educated gamers have always shown a stronger preference for games in the ancient and medieval periods. The less educated gamer leans toward more complex and large games, spends more hours playing them, and, for some reason, has a strong preference for World War I games. I never could figure that one out. I make no claims as to what any of this might mean. But the gamer who has been spending more years playing games is older and more educated, and owns a lot of games. He also shows a preference for Napoleonic games. Again, I don't know why. It's just what the correlation analysis revealed.

One of the earliest shocks we obtained from these surveys was from a casual inquiry about playing games solitaire. We ended up asking this question a number of times. Yet every time we asked whether people play the games with opponents or solitaire, we have consistently found that more than 50 percent of the games played are played solitaire. This is only partially accounted for by the fact that there aren't many gamers around. Another important reason is that players prefer to study the games without the hindrance of another person. In the 1990s, the number of games played solitaire exceeds 60 percent.

One of the principal reasons for all these surveys was to discover what type of game would be easiest to sell. Customers were continually asked about their preferences for complexity in games, their age, education, the number of years they'd been playing games, the number of hours they played a month, the number of games they owned, and the historical periods they preferred. In each of these characteristics, we found different preferences at each end of the spectrum. For example, the gamers who prefer the more complex games spend more hours playing them, favor the contemporary period, and like innovative, somewhat more complicated rules. The gamers who spend more hours playing the games own more games, and they too favor World War I games. And this may make some sense. The gamers who spend less time playing are usually older and more highly educated. The age-related items were most important as, during the 1980s, the younger gamers disappeared and the average age and education of gamers increased.

Some findings were obvious. Those gamers who own a greater number of games usually have been at it more years, spend more hours playing, favor more innovative games, and have a preference for pre-Napoleonic and World War I games. These two periods, by the way, are the ones that publishers paid the least attention to until quite recently.

Although we already knew that miniatures wargaming was a minority within the wargaming community, we were suprised to find that, at least in 1980, over 25 percent of wargamers had, at one time or another, tried miniatures. One sixth of those used naval miniatures, the rest land-warfare miniatures. Most eventually got out of the hobby because of the demands on one's time and pocketbook. Many former miniatures wargamers also noted that if you don't have the artistic skills to paint your own figures, you were at a disadvantage. Many also didn't want to spend that much of their gaming time only gaming tactical-level operations (which is what miniatures gaming basically is). Finally, most wargamers were in it largely for the information published games contained. In miniatures, it's largely roll your own. Do your own research and be your own game designer. Not everyone's cup of tea.

One of the enduring problems game designers and publishers are faced with is how complete a game's rules should be. There are so many possible things that can happen in a game that a truly comprehensive set of game rules could be (and some were) over 100 pages long. Fortunately, about two thirds of gamers said that game rules should be a set of instructions that completely covers all but the most extreme cases of detail, while about one in five felt the game rules should be an exact set of laws and regulations that covers any possible case without any need for player interpretation. This last is a nearly impossible task, even in computer games. The remainder of the respondees felt that game rules should be considerably looser. In response to these preferences, the level of game complexity had settled down to a large degree by the early 1980s. Before that, there had been some attempts at truly encyclopedic game rules. These herculean efforts were, and still are, admired. But they are also virtually unplayable. Computer-wargame designers have an additional problem in that the more features they put in the program, the more likely it is that the program will fail completely.

Following up on the questions of how extensive the game procedures should be, we also asked whether gamers preferred uniformity and standardization in the design of the games or more variety and innovation. Some 25 percent favored standardization or uniformity, 40 percent tended to prefer more variety and innovation, and 34 percent felt it depended too much on the game and could go either way. In any event, the trend has been more toward standardization, largely because this makes new games easier to get into. This has been a vital consideration for gamers who increasingly have less time to work with their wargames. It's a case of what you need versus what you want.

Despite the increasing time constraints among wargamers, they have managed to devote about the same amount of time to wargaming in the early 1990s (16 hours a month) as they did in the late 1970s (18 hours a month). One thing that has changed quite a bit is the time spent at any one session. In the 1970s, there were three or four sessions a month. Now there are a larger number of shorter sessions, indicative of the greater propensity to play solitaire or simply to study the games.

In terms of games owned, players tend to be collectors, much like book readers. The average gamer (somebody who's been playing five or six years) owns about 50 games. This usually includes a dozen or so games received in magazine format (at an average cost of $10) while the remaining games range in price from $10 to $100. This means an investment of several thousand dollars. The average gamer spends several hundred dollars a year on his hobby. On average a gamer will buy about a dozen games a year. Hobbyists tend to buy most of their games in stores. Even though many of the publishers go out of their way to make the games available by direct mail at more favorable prices, there is still a lot of impulse buying.

Which brings us to another subject. Why do people buy the games? We once ran a survey listing 28 reasons that someone would have for buying a game. We asked respondents to rate each of these factors on a scale of 1 to 9. What follows are the results (with the average 1 to 9 rating).

**1.** The subject of the game (8.36). This was the single most important determinator in buying the game. This makes sense. People tend to be interested in the subject. When it comes to wargames, the gaming element is simply a means toward an end.

**2.** A firsthand examination of the components (6.56). This is probably one of the big reasons why most people buy their games in stores, where they can handle the game. But again, people are looking for what the game actually does and how it does it. And since games are so visual, this naturally leads to the requirement for actually looking at the game. In the last 10 years, publishers have noted that graphic presentation tends to count for more and more, although it's unlikely it will displace subject matter as the most important consideration.

**3.** Publisher's reputation (6.51). This is something unique. In book publishing, the customer doesn't pay much attention to the publisher's reputation. It's the author that counts. In game publishing, it's just the opposite. This may be because the publishing end of the hobby has not matured enough to have developed and recruited outstanding "authors." Some publishers did push noted designers for their games, but results were not definitive one way or the other. However, another and probably more pertinent reason is that the publishing process for games, unlike book publishing, is more of a team effort. It's more like making a movie where you have actors, directors, and that unsung hero, the film editor, who puts the movie together after everybody else has finished his work. In a relationship such as this, much depends on the ability of the publisher to bring all of these people together and to get them working most effectively. Gamers apparently sense this; thus, the publisher's reputation ranks highly.

**4.** A review in a gaming magazine (6.21). People are always looking for a good opinion, or any opinion. Although there aren't too many gaming magazines available with reviews, they are becoming more important.

**5.** The graphic design of the game (5.93). This is actually a subset of number 2, and is becoming more important. I attribute this to the greater cultural emphasis on graphics. The TV generation and all that.

**6.** The initial perception of the game system (5.88). This means the ability of the gamer to perceive how the game generally works and how well he likes the way that it works. This element affords the publisher the opportunity to do some effective advertising.

**7.** Price (5.87). The survey found that if apparent value in a game is given, there will be no trouble selling it as long as the apparent value matches the price. This, however, represents a problem, since some combinations of game components, which cost the same as other combinations, will actually be perceived as more of a game. An example is the Quadrigames concept. The Quadrigame was first introduced by SPI in 1975 and was a set of four relatively small games in the same historical period that generally sold for the same price as a single but larger game. Now to a gamer, that seemed like a pretty good deal. That means the invididual games are costing him one quarter what one larger game would cost. However, it's actually cheaper to manufacture a Quadrigame (which has the equivalent of two full-size game maps and other components, making it it the equivalent of two ''normal''-size games) than to manufacture many other games that appear to be smaller. Because of the use of special components or colors, the ''smaller'' game costs more to manufacture. And even though the ''smaller'' game costs more to make, we generally have to charge less, since experience has shown that it is the perceived value that will weigh most heavily in a gamer's buying decision. In the final analysis, however, price is number 7 on this list of factors that are considered when deciding to buy a particular game. Quadrigames reappeared on the market in the early 1990s, as publishers noted that out-of-print ones were being resold at even-higher premiums.

In fact, on the ratings given so far, the subject of a game is by far the overwhelming determinant. I suspect that it is that factor plus a favorable score on three or four of the other top ten elements that will prompt a person to purchase a particular game. The remaining factors in making the buying decision are:

**8.** Description of a game in an ad (5.76).

**9.** The physical size of the game (5.71).

**10.** Whether the game is part of a ''family'' of similar games with which the gamer is already familiar (5.67).

**11.** The opinion of a friend (5.67).

**12.** Editorial description of the game-designing process in a magazine (5.67).

**13.** The availability of the game by mail order (5.23).

**14.** The reputation of the designer and/or developer (5.17).

**15.** The scale of the game (5.14).

**16.** The gamers' ratings of the game, published in a wide magazine survey of all publishers' games (5.0).

The other 12 elements were so far down as to be considered relatively unmeaningful in the buying decision.

Getting back to the most important determinator of the buying, we have period preference. Now this is the critical element among gamers, since the primary thing that separates gamers is the different period preferences. During the 1970s, certain trends became apparent. Generally speaking, the three major divisions in terms of periods are pre–20th century, 20th century, and fantasy and science fiction. In the early 1970s, about a third of gamers were into the pre–20th century period. Practically none were interested in fantasy and science-fiction games, because there weren't any. This left about two thirds interested in the 20th-century era. By 1980 this changed considerably. About 48 percent are interested in the pre–20th century age and 52 percent are interested in the 20th century. At the same time, 20 percent of the market for historical games had already disappeared into the fantasy and science-fiction camp. By 1990 there had been a massive shift, with 63 percent now favoring pre–20th century and 37 percent the 20th century. Actually, much of that swing occurred after the political changes of 1989. Without the Russian armed forces as a potential enemy, interest in "contemporary" subject all but evaporated. At this point, about half the potential wargame audience had made off to fantasy and science-fiction land.

We have arbitrarily (well, not that arbitrarily) divided the three major time divisions into nine periods. These are:

Ancient (Rome, Greece, Biblical, 3,000 B.C. to A.D. 600)
Dark Ages and Renaissance (600 to 1600)
Thirty Years' War and pre-Napoleonic (1600 to 1790)
Napoleonic (1790 to 1830)
Civil War/19th Century (1830 to 1900)
World War I (1900 to 1930)
World War II (1930 to 1945)
Modern (1945 to the present)
Fantasy and Science Fiction.

In the early 1970s, 20 percent of gamers were mainly interested in World War II. By 1980 this was only 18 percent, and by 1990 it was only 8 percent. World War II was fast fading from memory. In the early 1970s, about 28 percent were interested in contemporary subjects; by 1980 it was only 18 percent. This was a curious result, as games on contemporary subjects were always good sellers. What was apparently happening was that a lot of nonwargamers were buying the contemporary subject games (and probably going into shock when they broke them open). The regular wargamers were, even then, developing more interest in pre–20th century subjects. Through the 1980s, perhaps because of the political climate, wargamers expressed more interest in contemporary wargames. It went up to nearly 30 percent by 1989, after which it rapidly declined to 21 percent because of the demise of the Big Red Wolf. NATO/Warsaw Pact games passed into the realm of science fiction, for the moment, anyway.

This is something else to consider. Taste in games by period tends to vary from year to year. It goes in waves. For example, in the early 1970s, 18 percent of gamers were interested in the Civil War. In the mid-1970s, this declined to less than 10 percent. By 1980 it had increased to 16 percent and was growing. World War I was favored by about 5 percent of the gaming audience in the early 1970s and has stayed the same for the decade. Ten years later, it was up to 8 percent.

Plotting the ups and downs of period popularity over 20 years clearly shows that periods regularly wax and wane. Periods are relatively hot for a few years, then they decline before becoming popular once more. Generally, periods have an average, which varies a great deal from period to period. Some periods, in the long term, are more popular than others. The only dramatic change took place in the late 1980s, when, in effect, World War II finally ended with the collapse of Communism and Russian military power (conventional military power, at least). This knocked the legs out from the popularity of World War II and post–World War II games. However, post–World War II games still have the relatively high popularity that World War II enjoyed for over 40 years. You could almost hear the gears of history ratchet forward one notch.

Periodically, I have also asked gamers which periods they liked the least, for marketing surveys indicated that while gamers would surely buy games they liked the best, they were also inclined to try games from other periods, except for those periods they liked the least. As of the early 1990s, this is the lineup of periods liked the most and least.

| *Period* | *% Liked Best* | *% Liked Least* |
|---|---|---|
| **Ancient 3000 B.C.–A.D. 600** | **20** | **10** |
| **Dark Ages/Renaissance A.D. 600–1600** | **8** | **12** |
| **Thirty Years' War/Pre-Napoleonic 1600–1790** | **12** | **7** |
| **Napoleonic 1790–1830** | **8** | **5** |

| | | |
|---|---|---|
| **Civil War, 19th Century 1830–1900** | **15** | **11** |
| **World War I 1900–1930** | **8** | **8** |
| **World War II 1930–45** | **8** | **8** |
| **Modern 1945–Present** | **13** | **11** |
| **Future/Hypothetical** | **8** | **28** |

And so it goes.

A number of things are happening here with these changes in period preference. First of all, older gamers are quite naturally improving their education and getting more interested in more historical periods they haven't covered yet, namely, the older epochs. At the same time, younger gamers (particularly teenagers) who previously were more interested in World War II and modern topics (if only because they were more recognizable) are tending more toward fantasy and science fiction. However, because life is a circular process, these younger gamers are also coming right back to the older eras as their education increases. In fact, there's a tremendous overlap and interest between fantasy and science-fiction gamers and players interested in the medieval period. This is not as strange as it may appear, since much fantasy lore derives from medieval times (heroes in shining armor wielding blood-drenched swords while grasping the damsel in distress).

For a gamer to stay in the hobby for a long time, he must find the time to play, or at least to work with the games as information sources. The younger hobbyists, the ones who have been in it for less than two years, generally play less than 12 hours a month, whereas those who have been at it for nine years or more will play 16 or 17 hours a month. An older person often has less time than, say, a student. He has more obligations, more pressures. Sticking with wargames requires a considerably greater investment of energy from an older person. This is probably one of the reasons why the games are so "serious," that is, demanding considerable fidelity to accuracy and research and presentation of the information. The games are addressed to people who are either already quite well established in responsible positions, or are on their way to that. Sometimes appearances can be deceiving. I once received a letter from a fellow in Tennessee who had visited me at SPI a month earlier. As was our custom (it was my turn), I had taken him on a tour of the place. He was dressed somewhat as I was, in shabby jeans and workshirt, and made reference to that fact in his letter thanking me for the tour and the hospitality. He had an M.D. after his name and noted that I should drop in and say hello whenever I was in Tennessee, although I'd probably never have occasion to use his services as a gynecologist. This is fairly typical. Many gamers are doctors. I often wonder where they find the time to play.

Speaking of occupations, wargamers cover a fairly wide range. Some of the things hobbyists do for a living are rather understandable. Many players are still students. Nearly all age 22 and under are. Taking the entire gamer population, about 20 percent are in the military (including Department of Defense and CIA civilians). Some 5 percent are officers in combat units, another 4 percent are NCOs and troops in combat units. That means about 10 percent of hobbyists are in combat-related jobs in the armed forces. There's always been a lot of commercial wargaming going on in the military. That's where I first encountered wargames in the early 1960s. When I ended up with an artillery battalion in Korea (1963–64), there were about half a dozen enlisted wargamers and three officer wargamers in that one battalion. While at SPI, I was visited by a player who was an engineering officer on a nuclear submarine. He commented that one of their favorite games while they were on very long cruises was *Frigate*, a game of naval warfare during the 19th century. During the 1980s, many of these military people began designing commercial wargames. Gary ("Mo") Morgan, an F-16 pilot, has designed several games for Avalon Hill and one on the 1812 Battle of Borodino for *S&T*. Larry Bond, a civilian analyst for the navy, designed *Harpoon.* Several CIA analysts have had wargames published, usually on pre–20th century (or pre–World War II) subjects. There are others, and there will be more. Wargamers in the military are a growing group, and they are the one group that has contributed much to keeping the hobby alive. I should note, however, that at West Point, through the 1980s, an increasing number of cadets prefer *Dungeons and Dragons* and other role-playing games. What this means for the future of the U.S. military, we can only speculate about.

Some 5 percent of gamers are employed by the government. This includes a lot of people in the Foreign Service, CIA, various national-security agencies, and similar operations. They consider gaming to be something of a professionally related activity, which, for many of them, it is. About 10 percent of gamers are in blue-collar occupations, ranging from an in-flight food-service worker and machine operator to farm workers and others. This is an interesting group, which includes a midwestern pig farmer who has what is arguably one of the largest collections of World War II Order of Battle information in the world. I've known several bus drivers who were keen and knowledgeable wargamers, including one who had graduated from an Ivy League college but preferred driving a bus and studying history. A third of wargamers are in what we call professional occupations. These include systems analysts (3 percent), programmers (2 percent), corporate officers (1 percent), lawyers (3 percent), clergy (1 percent), doctors (1 percent), scientists (5 percent), and so on. I'm never bored when I get into a conversation with hobbyists. Many of them are doing more interesting things than I am.

Given their income and training, it's not surprising that wargamers were among the first to get personal computers; by the early 1990s, most had them, and the majority of those played computer wargames. Several have commented that computer wargames have made the hobby a bit more respectable. This raises the long-standing problem of wargames being considered mere games and their users mere gamers. Such is not the case, as we have seen, but the label sticks and causes discomfort at times. Playing a game on a computer has a somewhat more respectable aura about it, especially since the data displayed on wargame screens looks a lot like the professional wargames the Pentagon creates.

By most definitions, wargamers are a select group, above average in education, income, and, especially, diligence. Wargames are not easy to master. Unique mental skills are required to deal with all that goes on in a wargame and make the game work. Even computer wargames are considered a cut above average in complexity compared to most other games. But what most distinguishes wargamers is their desire to explore the past and find out not only what has happened, but details of why and how it happened. Considering the continued popularity of games on contemporary subjects, wargamers also want to apply those historical insights to the present and future. More experienced wargamers also learn that much of what is generally considered war is not combat, but politics, sociology, and everything that defines the human experience. Which is why many wargamers wince at the term "wargames" and prefer to call them what they really are: historical simulations.

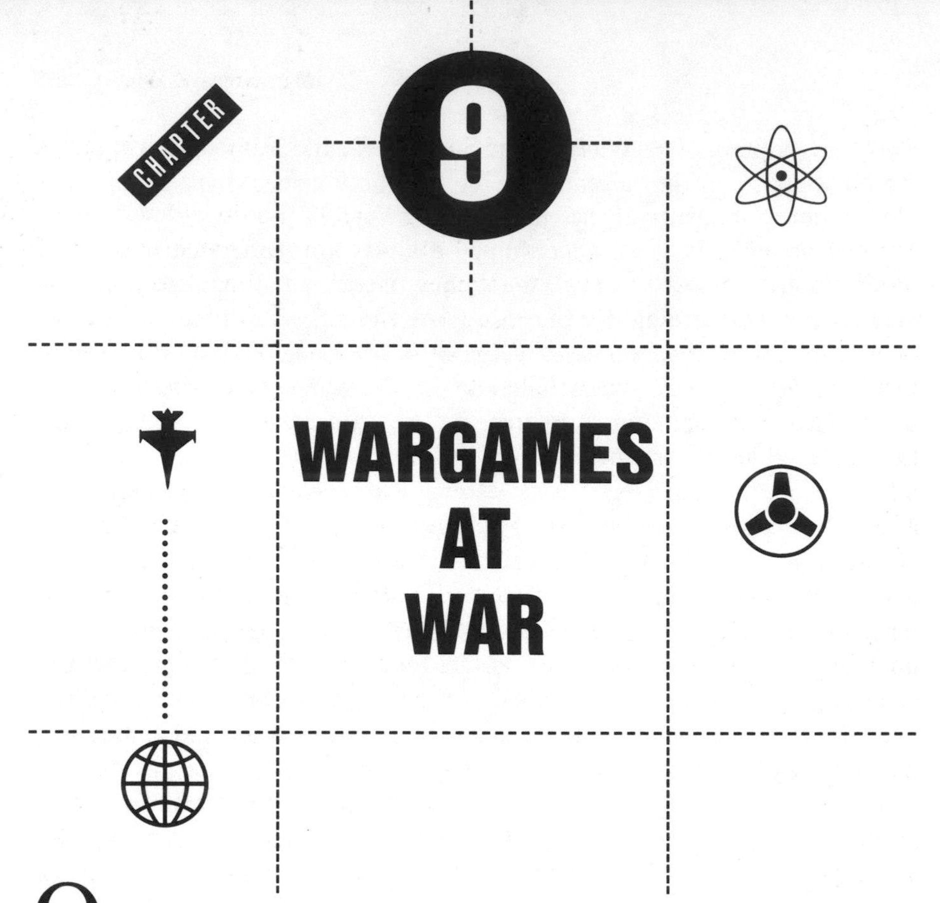


On August 2, 1990, as Iraq was invading Kuwait, the high command of the American armed forces in Washington, D.C., swung into action. One of the first things the Pentagon officials did was to wargame out the unfolding situation. All of the Pentagon's computerized wargames were too slow off the mark for this job, so later that day, the first wargame used at the Pentagon was a commercial one. With all the billions spent on computerized wargames since 1945, America's most efficient military operation in this century was initially planned using a game (*Gulf Strike*) that could be bought by anyone in most hobby stores for under $50.

During World War II, the German Army regularly wargamed operations in much the same way. The manual wargames used by the Germans were very similar in style to current manual games (although the Germans considered their games military secrets and not available to civilians). When the Allies invaded France on June 6, 1944, the Germans were in the middle of a wargame dealing with just such a possibility. As reality had overtaken the game's hypothetical premise, the German commander ordered the game to proceed, not as a game but as a command tool. Wargaming had been a common practice in Germany for a century before the Nazis came along. The British, Russians, and Japanese also wargamed every major operation. In the United States, only the Navy used wargames for planning.

However, for many years after World War II, traditional wargames got little respect from the professional military wargamers because these older, history-based, manual wargames were not considered precise enough for their needs. The development and application of computers, operations analysis (OR), and systems analysis during World War II firmly planted the idea that simulations of war had to, and could, provide precise and unambiguous answers. These professionals ignored the fact that history-based wargames, despite being relatively imprecise, had usually been accurate enough to be useful. But, as the decades rolled by after 1945, a lot of the OR-style wargamers began to realize that absolute accuracy was not only difficult to attain, but often unnecessary, much less match the reliability and accuracy of the older-style wargames. Finally in the 1980s, as even the slow learners realized that the precise and unambiguous approach could not be realized, chaos theory emerged and (without getting into the technical aspects of *that*) gave the high-precision crowd a scientific reason for admitting that the older wargaming techniques did indeed work.

Several military wargamers summed up the problem very succinctly by pointing out that there is a trade-off between the accuracy of your wargame and how much time and resources you have available. The relationships for these various analyses, performed by an experienced analyst, can be expressed thus:

- A one-hour, +/− 20 percent accurate solution can be done on a piece of paper with a calculator. For example, trying to solve the problem of how long it would take an armored division to travel a distance and fight a battle and what resources would be needed and what losses would be incurred. The method described here would give you an answer either 20 percent higher or lower than reality. Well, that may be a little optimistic, but it's in the ballpark and makes the point.
- A one-day, +/−10 percent accurate solution can be done on a spreadsheet program running on a personal computer. This approach uses much more data, calculation, and accuracy.
- A one-week, +/−5 percent accurate solution can be done on a spreadsheet with current data gathered from many sources and elaborate data handling and calculation routines created in the spreadsheet. Much of the time would be spent gathering the information, always a necessary and tedious task for producing wargames.
- A one-month, +/−1 percent accurate solution can be used with linear programming or other analytic tools (also on a personal computer)

and even more data. Linear programming can also be used in conjunction with a spreadsheet model (a technique I favor).

Until recently, professional wargaming tended to go after the less than 1 percent deviation in accuracy and consume years and millions of dollars in the process. Indeed, the process usually overcame the search for a solution, often leading to a lost, and failed, project. Commercial manual wargames can generally achieve a "5 percent solution." Put a wargame on a PC, and you get a nearly 1 percent solution. But it took the military a while to catch on to this cheaper solution sitting on the shelves of a local game store.

## WARGAMES, MODELS, AND SIMULATIONS ✱✱

One problem that has caused some confusion within the military wargaming community is the use of the terms "wargame," "model," and "simulation." It causes even more problems among civilians, who also hear the term "conflict simulation" tossed about. Wargames are, like most games, also models and simulations of real-life events. The three terms are commonly (and incorrectly) used interchangeably, but each term means quite something different to the military wargamer. Wargames are usually simpler than models and simulations because, as the names imply, a wargame is something of a competitive game that is played, while a model is a more detailed representation of a specific military event. A model duplicates a function in great detail and exactitude. A simulation is a model, or collection of models, that can be more easily manipulated to test "what if" questions. A simulation is a model that can move in many different directions. A wargame is a playable simulation. A conflict simulation is another name for wargame, one that leaves out the two unsavory terms "war" and "game."

The U.S. military maintains an inventory of nearly 600 different wargames, simulations, and models. Some of these are quite arcane, especially those dealing with various aspects of electronic warfare. A lot of them are over 30 years old, including *ATLAS*, one of the oldest and most ill-conceived models ever put to use. Wargames are a minority among this 600, with most being models, plus a lesser number of simulations. Most of these items are used for examining very narrow issues. Typically, these are models developed to support some engineering need. There is a tremendous need for models of this type, as any competent engineer will want to model a new system thoroughly before handing it over to the troops. Similar models are also used as training devices for the troops using the new weapons systems. Unfortunately, this plethora of engineering-type models has, quite naturally, heavily influenced the design of simulations and wargames within the military. This is only natural. With so many models in use, it's hard to avoid using models as your prototype for developing simulations and wargames.

These 600 items were largely created by civilian contractors (the "beltway bandits") who draw their staff from academia and retired military personnel; the rest are created by active-duty troops or civil servants. In the last 10 years, about 5 percent of all military officers have received some OR or wargames training, and increasingly these military gamers are creating the military's wargames without assistance from the beltway bandits.

Operations research and wargamers (they tend to be clumped together) have their own organizations within the military, led by an assistant secretary of defense for systems analysis. Each of the services has one large and several smaller organizations devoted largely to OR and wargames work. The National Defense University (for all services, in Washington) and each of the services' war colleges have active wargaming operations.

The Army has the CAA (Concepts Analysis Agency, in the D.C. area), which provides wargame analysis to support new doctrine, weapons, and tactics as well as justification for budget requests. The army also has its TRADOC (Training and Doctrine Command) Headquarters in Fort Monroe, Virginia, where a lot of systems analysis is done. The air force has a Studies and Analysis operation in Air Force Headquarters at the Pentagon, providing the same services as CAA, but with less emphasis on wargaming. The navy has its Center for Naval Analysis in the D.C. area, which, like the air force, does more analysis than wargaming.

Most of the wargaming is done outside Washington. The army has a major wargaming operation at Fort Leavenworth, as well as several TRADOC Analysis Centers (TRAC). White Sands, New Mexico (where missiles are developed), is another analysis center. The bulk of the detailed analysis (and a lot of the gaming) is done at the branch schools (Armor, Infantry, Artillery, etc.). Fort Lee, Virginia (the logistics school) is where the army does most of its OR training and some wargaming. Weapons-effects data is obtained from the Aberdeen (Maryland) Ballistic Research Lab and other facilities at Aberdeen. Note that much of the data coming out of Aberdeen over the years was later found to be inaccurate, either because the information request was not understood, or the wrong data was sent and the inaccuracy was never discovered until much later (if at all).

The navy does most of its wargaming at its Newport News (Rhode Island) war college. The navy operation is the oldest, largest, and arguably the most effective in the world. The Naval Postgraduate School (Monterey, California) is the Department of Defense center for degree study of OR.

The air force has a relatively new wargaming center at its war college in Mississippi.

The Joint Chiefs of Staff Group (at the Pentagon) does analysis for budget requirements and planning, as well as some wargaming.

Washington, D.C.–based contractors tend to supply the bulk of games and simulators used. Because there are so many experienced wargamers in uniform, wargaming projects have been popping up all over the place. Few if any of these games get entered into the official inventory of nearly 600 Department of Defense wargames, models, and simulations. In practice, only a few dozen of the 600 official wargames are widely used. Through the 1980s, though, a flood of new wargames came into official use, largely inspired by the commercial wargames available in stores and a recognition of a need for new wargames that performed better than the old ones.

## THE MILITARY EXPERIENCE WITH WARGAMES ✱✱

Wargames in the military have a long history, as the concept of working out battles ahead of time struck many ancient soldiers as an easy way to gain an advantage. Chess was an ancient wargame developed to train apprentice commanders on the finer points of battlefield operations as they then existed. In the 1700s and 1800s, chess developed into more complex wargames that most gamers today would recognize. But it was only in the last 30 years that hobby wargames came to infiltrate the previously computer-dominated regions of professional military gaming.

Military wargaming had been quite popular and effective from the early 19th century through World War II. These wargames were quite similar, and sometimes identical, to the wargames described in this book. But several things happened after World War II that took wargaming down a blind alley. These were:

• The success of operations-research techniques in supporting military operations during World War II.

• An identification of wargames with the Germans, the major symbol of evil during World War II and regular users of wargames throughout the war.

• A move away from the use of historical study to formulate future plans and policy (especially in the army and air force).

• Ignorance of the fact that Russia still wargamed under a thick blanket of secrecy. The United States, which dominated the rest of the world's military thought, had lost the habit of wargaming. The U.S. Navy still wargamed, but the army and air force dominated the U.S. armed forces, and neither of these services had any more interest in wargaming.

The net result was that the traditional, history-based manual wargame was replaced by computer-driven simulations of current and future events, designed using operations-research techniques. This caused four unfortunate results. First, operations research was not capable of accurately modeling all the chaotic events that take place in combat. This has only recently been acknowledged as a serious problem and is finally being addressed. Second, the Cold War created an atmosphere in the Pentagon that made dispassionate analysis of U.S. and Russian forces difficult and usually impossible for political reasons. The Russians were the enemy, and enormous U.S. defense expenditures could only be justified if the Russians were always portrayed as an awesome threat. Realistic, history-based wargames would show Russian limitations, and this was not acceptable.

Third, the discarding of military history as a tool made it impossible to catch the errors. The OR-based wargames created an artificial world where the possible outcomes fit the preconceived ideas of senior military and government leaders. With the historical approach, you could get a reality check from recent experience. The OR-based wargames soon lost all credibility with the military, and much of the civilian leadership. And last, the problems noted above caused wargaming's original purpose, officer training for combat command, to fall into disuse. Instead, the "models and simulations" were given the job of finding out what future wars would require in terms of weapons and ammunition. In other words, "wargaming" was reduced to a logistics-support function. These mainframe-based "wargames" were also used to justify most new weapons systems.The operations-research approach had turned into a Frankenstein monster.

Operations research (OR) is a 20th-century technique combining science and common sense. It sounds arcane and intimidating, but the concept is basically very simple. What OR tries to do is get to the bottom of a problem and implement a solution. For example, an OR practitioner would improve the manufacture of an item (like automobiles) by carefully examining all the steps and components that go into making a car and then systematically measuring and testing all of those steps and procedures to find which ones could be better or more efficient. Widely used by industry to fine-tune its operations, OR got its first military exposure during World War II. During the war, military operations became a great deal more technical, and OR was a technique that was in the right place at the right time. OR was widely used in logistics, manufacturing, in the battle against German submarines, and by the air forces.

Ironically, many wargaming techniques are the same as or similar to those used in operations research. Wargames are not as scientifically rigorous as OR, but that is largely because the first wargame designers realized that attempts at rigorous simulation were doomed to failure due to the complex nature of warfare. The earliest wargames also had no computers to assist with the bookkeeping. Similar complexities (revealed by chaos mathematics) exist in the earth's weather, with the result that forecasts for more than 48 hours in the future become completely unreliable. More powerful computers have gradually made longer-range forecasts practical. Yet warfare is even more changeable than the weather, and there is less plentiful (and less accurate) historical data available on warfare. But wargaming does have one advantage over weather forecasting, and it is summed up in the ancient military saying, "It's not a matter of who's better, but who's worse." Victory goes to the side that has more advantages, and playing out the maneuvers and uncertainty of warfare in a crude wargame tends to give the wargamer the edge.

Part of this "wargamer's edge" derives from the analysis the wargamer must perform to play the game or, as many officers do, create his own on the spot. A wargame is a very organized look at a military situation, because a wargame must have precise rules and accurate information in order for it to portray a military situation with any degree of success. Because battles occur with much less frequency than the weather, most soldiers have only a vague idea of what combat will be like. They are much more likely to be familiar with, and prepared for, say, a blizzard. Wargames give the soldier a hint of what combat is like, and this provides an edge that is often crucial. Yet many military users of wargames are unaware that they are using OR techniques (albeit usually the simpler ones) to create and use their games. Many OR professionals *are* aware of this connection, which is why many in the OR field, particularly those with military experience, have taken to wargames in the past 20 years.

OR professionals are accustomed to examining military events from a very narrow perspective. Wargames enable them to see combat situations in a broader context. Wargames require analysis of terrain, Order of Battle, losses, and many other factors. Those who have used historical wargames also become familiar with the techniques used to connect real events with the same events depicted in a wargame.

## WARGAMING AND THE PROFESSIONAL WARRIORS ✱✱

In the 1970s, the commercial wargames began to have an influence on the design and use of professional wargames. At first there was an indirect influence of the hobby games, as the troops bought and played them, which had a subtle and enduring impact. By the end of the 1970s, the military wargamers (users first, then the designers) found that the simpler techniques used by commercial wargames were more effective at simulating warfare than their own highly complex and heavily computerized efforts. A common criticism directed toward professional wargames during the

1970s was insufficient attention to historical experience in order to validate the military simulations and models.

These rumblings within the professional wargaming community were one of the causes of the 1977 Leesburg conference, the first gathering of all the major forces in military wargaming. Two others were held, one in 1985 and another in 1991. I was invited to all three, but the first invitation was a clear sign that things were changing, as I was an outsider. It was obvious that the winds of change were blowing strong when Andrew Marshall, a senior official of OSD (Office of the Secretary of Defense) and one of the key sources of funding for professional wargames, got up in front of the assembled multitude and stated bluntly, ''You people have never given me anything I can use.''

When my turn came to speak, I pointed out that what was needed was a wargame the commander could sit down with and operate himself. Having the ultimate user of wargame results actually operate the wargame would save a lot of time, get much better results, and eliminate a lot of confusion. It would also enable the commander to experiment with options that he might be reluctant to try through his staff (because the idea, or the results, might prove embarrassing). This last point is important, as the sociology of senior command makes it difficult for a commander to appear ignorant of anything or capable of doing something stupid, especially in front of subordinates.

By the late 1980s, the technology existed for such ''commander games'' (it didn't in 1977), and today you can pretty much play the ''commander game'' with many current commercial computer wargames. This spotlighted a major deficiency of military wargames, what commercial wargamers call the ''interface.'' Creating a commercial wargame, whether paper or computer-based, requires that up to half (or more) of the effort (and budget) on the project be devoted to how the game looks and how easy the game is to use. If a game looks ugly and is too difficult to use, it won't be used and, more important, won't be bought. Because of a lack of commercial pressures, the military games never had very user-friendly interfaces. Moreover, the principal users of military games were the people who created them, or computer professionals accustomed to arcane interfaces. There was never any big incentive to develop efficient interfaces.

At Leesburg and subsequently, I urged the military wargamers to buy the interface (and wargame) technology from commercial publishers. I had already sounded out several publishers, and most were quite eager to work with the military. In time some military organizations did obtain technology from commercial wargame publishers, but only on a limited basis.The resistance to this approach was threefold: First, commercial publishers were not accustomed to all the paperwork and bureaucracy required to work as a "government contractor." Second, many government agencies were not experienced at going out and contacting firms that were not actively seeking government work. Third, the existing government suppliers of wargame technology did what they could to discourage the purchase of these "toys" (commercial wargames), as the "toys" were a lot cheaper and more competitve than the multimillion-dollar military wargame projects that kept so many defense consultants (and many government employees) comfortably employed.

Even before Leesburg, I had been approached to help integrate commercial wargames into the military. Yet this episode demonstrated that helping the military develop wargames was not without its pitfalls. In 1972, while running a wargame-publishing company, I was approached by a group of young officers at the Army Infantry School, who asked me to design and publish a game on a contemporary subject. So I designed *Red Star/White Star*, which went on to become a great success after its publication in 1973. This attracted the attention of more senior officers in the army who encouraged the development of a commercial-type wargame for use in training troops. It took a few years, but eventually, with the help of a $25,000 development contract from the army, we produced the game, called *Firefight*. The army initially accepted, but ultimately dropped, *Firefight* for training purposes because it was "too complicated." But it did get a couple of thousand copies circulating among combat units. However, I estimate that we sold more copies of *Firefight* to soldiers than the army distributed, since our contract allowed us to publish it for the general public. The game was a best-seller, and as of early 1992, was still in print.

The *Firefight* project was an interesting example of how the military (or, I should say, the large peacetime military organization) operates. All reports I got back from evaluators of *Firefight* recommended it. I spoke to one fellow who headed the evaluation team for the Fifth Corps in Germany; he said that his team's review was highly favorable. The army even went to the expense of rewriting the rules in a much more expanded and illustrated format. This undoubtedly cost much more than it cost us originally to do the game. I concluded that for most training purposes, the game probably was too complicated for many potential users (senior NCOs and junior officers). *Firefight* required too much work to play. It should have used a computer. But computers were not yet generally available, and *Firefight* was. However, many of those troops who were able to get a copy found it useful. At least it showed them what a manual wargame was and what it could, and could not, do for them.

Manual wargames require work, more work than a nongamer who does not need to use the games professionally might want to invest. I have repeated this observation to any number of military organizations when questioned about the use of wargames for training. The obvious solution is to use microcomputers. In the late 1970s, I was talking about the future, but now the future has been with us for a few years. The military has been using microcomputer-based wargames increasingly through the 1980s and with increasing success, although never to the PC's full potential.

Yet even before personal computers became common and powerful, there were examples of military wargamers successfully combining existing computer technology and common sense to produce accurate and useful wargames. In 1976, Colonel Ray Macedonia asked me to come down to the Army War College and help him reintroduce the use of wargaming to train the army's future generals. This effort also resulted in the design of the *McClintic Theater Model* (*MTM*), which was based on a manual game I had designed a few years earlier. *MTM* became popular and widely used during the 1980s. In 1980 I met with Andrew Marshall and agreed to design for him a global simulation called the *Strategic Analysis Simulation* (*SAS*) for the National Defense University. Mark Herman completed and implemented the game, and it's still being used today.

During the late 1970s, the army and navy began using a number of wargames based in part on commercial wargame designs or techniques. These included *Dunn-Kempf, Pegasus*, and *First Battle* for the army, and *NAVTAG* and *NEWS* for the navy. Many of these were computer-assisted, while *Dunn-Kempf* (named after the two army officers who designed it) used miniatures.

I was also called upon to do a lot of lecturing to military audiences. My main theme was that commercial wargame techniques were readily available, quite easy to implement, and did not require a lot of money. I pointed out that there were already a lot of wargamers within the military, and if you put the word out, they would present themselves, ready to design the games the military needed. These military wargamers soon came forward, in the thousands. Not only the troops themselves, but often their teenage sons who were also wargamers. Ray Macedonia used such young wargamers in developing games at the Army War College.

Over the years, I received many letters and verbal reports of units implementing their own wargames, reflecting their local situation, as a supplement for field training. Eventually, several of these military wargamers made it to the big time, having their designs published. The most notable of these were Air Force F-16 pilot Gary Morgan (author of *Tac Air, Flight Leader, Borodino,* and *Jihad*), army officer Bill Gibbs (*Ranger, Main Battle Area, AirLand Battle*) and navy civilian analyst Larry Bond (*Harpoon*, in manual and computer versions). Gary Morgan's experience was most interesting. His first two published designs (*Tac Air* and *Flight Leader*) began as official air-force wargames called *FEBA* and *Check Six*.

This was unusual, although many other manual wargames designed by uniformed wargamers for professional use were of commercial quality.

A lot of the wargaming activity in the military after 1975 was a result of the soul-searching and restructuring that followed the trauma of Vietnam. It wasn't until the U.S. Air Force and Army made mincemeat of the Iraqi armed forces in 1991 that most people realized how drastically the American armed forces had transformed themselves in the previous 16 years. One of the primary engines of that transformation was wargaming, the kind of wargaming covered in this book. After 1975, new ideas were welcome in the American military. If a new idea worked, it was widely adopted. That's what happened with wargames.

## PROFESSIONAL WARGAMES AND MILITARY DECISION MAKING ✱✱

A lot of American military planning and wargaming problems went right back to 1945, and the end of World War II. The first problem was the introduction of nuclear weapons and strategic bombers, which left the army without a mission. As nuclear weapons increased in importance, the navy and air force became the principal services, with the army reduced to the role of a "trip wire" force. If any army units scattered about the globe got attacked, the plan was to throw nuclear weapons at the problem. Even the Korean War, in which the United States refrained from using nuclear weapons, did not change this attitude. Right through Vietnam, where nuclear weapons were again not used, the army was struggling to find a "mission," some clear statement of what the army was supposed to be doing. Although the navy and air force were getting most of the money, the army was traditionally the senior service and, despite everything else, acknowledged as the nation's final line of defense against any enemy threat. Yet the official policy was that nuclear weapons would defend the nation, even though the army got called out time and again after 1945 whenever there was a problem.

The Department of Defense either had no wargames with which to explore this situation, or it had inadequate ones. This situation persisted until the 1980s. Before the 1980s, there were numerous attempts to use wargames to solve the Army's "mission problem." Vietnam, and realization that nukes were not much good against guerrillas, produced one of the more successful attempts. This was a five-year model-building effort that produced a pretty good "low-intensity conflict" (LIC) model in 1971. Like most operations research–based efforts, though, it did not give all the potential "what ifs" a real workout. Users of commercial wargamers learned to expect the unexpected, but Department of Defense wargames tried as much as possible to eliminate the unexpected. Moreover, the LIC model required a few months for several people to learn how to use it before it could be run. By the time the LIC was completed, the U.S. withdrawal from Vietnam was in high gear, and attention was shifting back to Europe and the concept of an increasing Warsaw Pact threat. The military decided it was unable to use combat modeling and simulation to solve the Vietnam dilemma, or similar ones. Still without decent wargames, it turned yet another set of OR tools toward wargaming the Russian threat in Europe.

While LIC was something of a high point in military "black box" wargames, there were few others. One of the principal combat models for wargaming operations against the Russians was *ATLAS*. This was a resource ("how much ammo, troops, etc., do we need?") model developed by the Army War College and IDA (Institute for Defense Analysis, a DoD think tank) in the 1960s for resource allocation in a European war that modeled air and land operations. It did so in a very primitive "piston" (two sides pushing at one another) style of modeling. Initially, *ATLAS* had look-up tables that a human could understand, but later models got so abstruse that users had no idea how results were achieved. Another very complex model was *IDAHEX*, which was a replacement for *ATLAS*. In that case, the new model was ordered not so much to correct *ATLAS*'s failures, but to support MBFR (Mutually Balanced Force Reductions, an early 1970s attempt at reducing conventional force levels in Europe by 50 percent, at the behest of the Congress). MBFR negotiations required a model that would show the trade-offs between air and ground resources (and naval too, as the Russians considered U.S. air and naval resources major factors). The MBFR talks never accomplished anything, and neither did *IDAHEX*. There were a host of reasons for this lack of performance on the modeling end. The older "piston"-type wargames like *ATLAS* and *IDAHEX* could not deal with maneuver warfare. Naturally, Russian doctrine then, as now, was premised on a lot of maneuver. These models tended to have great detail for one service and hardly any for the others. The major problem was that the key people in OR gaming were scientists who were trying to use the scientific method to solve all military problems. Before, military people had used history to do a top-down analysis and solution. Scientists used a bottom-up approach. A good example of this is the "Monte Carlo duel," in which a model of one tank fighting another tank is meticulously modeled. What is missed is that on the battlefield there are few strictly one-on-one engagements. Rather, there are many tanks, and infantry, interacting in a quite complex and chaotic fashion. The strictly technical method was rigid and scientifically complete, but unable to simulate the battlefield.

The Vietnam experience finally caused a lot of reform-minded officers to come out and demand a defined doctrine, a well-thought-out and practical plan for what the army was to do in combat. There had been none before other than to station army troops in Europe, backed by MAD (Mutually Assured Destruction, with nukes). This was fundamentally vague and basically meant a nuclear holocaust if it was ever used. Up through the 1970s, no thought was given to ''what ifs'' and what would happen in a rapidly changing battlefield situation. Something in the way the military operated had to change.

Three things did change in the early and mid-1970s. First, less attention was paid to the traditional models like *ATLAS*. Second, all those officers who had been playing commercial wargames since the 1960s were now moving up in the ranks and were able to shift attention to these history-based games and models because, increasingly, wargamers (or those who knew what a wargame was) were making the decisions. Third, and most important, the study of history returned to fashion. Aside from the reintroduction of wargaming at the Army War College and the wider use of wargames in general, these changes led to a new combat doctrine for the army in 1976: ''Active Defense.'' While a new look at World War II battles influenced this new development, the Israeli success in Israel's 1967 and 1973 wars also was a big factor. The 1973 war in particular convinced many in the army that a high-tech army could fight outnumbered and win. Active Defense was a departure from the nominal post–World War II doctrine of attrition: of going head-to-head with an opponent and grinding him down. Active Defense borrowed from the German experience fighting the Russians in World War II. Although the Germans lost the war, on the battlefield they generally inflicted far more casualties than they received. The Russians, like the other Allies, beat the Germans with superior numbers. But the Germans did demonstrate that you could fight battles outnumbered and win. As NATO forces were constantly outnumbered by Russian forces in Central Europe, they had no choice but to develop a doctrine that could do what the Germans had done, and do it a bit better.

At the same time Active Defense was being introduced as the new way of doing things, wargaming was demonstrating that the new doctrine wouldn't work without the cooperation of the air force. Getting the air force to cooperate with the army in developing and using a joint doctrine was no simple task. But a combination of diplomacy and wargaming did the job, and by the early 1980s Active Defense had turned into AirLand battle. The U.S. Army had got good results from the air force in World War II because, at that time, the air force was still part of the army. But after World War II, the army air units became a separate service. Because of this, for thirty years, the army and air force drifted apart in terms of doctrine. AirLand battle brought them back together.

During the 1980s, wargaming was used throughout the army, and to a lesser extent in the air force. Wargaming was slowly becoming established as an official way of thinking and planning. Wargames (including modified off-the-shelf commercial games) were used to clarify and focus warplans for Middle Eastern contingencies in the late 1970s and early 1980s. All could see the results of these changes in the 1991 Gulf War.

The Army War College (at Carlisle Barracks, in Pennsylvania) was established in 1901, in response to the disastrous lack of planning and coordination during the 1898 Spanish-American War. Wargaming was there at the beginning, but never really caught on. Wargaming disappeared at the war college before World War II, and a wargaming department was not reestablished until 1981. This department lasted only a few years before disappearing in yet another school reorganization and reemerging in 1988 as the Department for Strategic Wargaming. Wargaming survived, this time, because there were enough convinced wargamers around.

How wargaming returned to Carlisle is a typical tale of how a few individuals in a large bureaucracy can make a big difference. In the mid-1970s, the commander of the Army's Training and Doctrine Command (TRADOC), being a history buff and noting the mischief I was involved in with wargaming officers at the Army Infantry School, looked around the Army War College and selected recent graduate Colonel Raymond Macedonia (a paratrooper with OR training who had worked on OR-type models at the Pentagon) and basically told him to get wargaming going once more at the war college. Ray asked around about who might know something about wargaming, and my name came up. I then got a phone call from Ray asking if I would volunteer to help reestablish wargaming at the War College. I said yes, and proceeded to regularly commute from New York City to central Pennsylvania for the next seven years. In 1978 the then–chief of staff of the army began using the wargaming facilities at the War College for planning future army operations. One of the most amazing things Ray Macedonia did at the War College was to build a computerized wargame that worked, and that would be useful to the students. He pumped me for ideas, concepts, advice, information, and prototypes, and then turned all that over to programmer Fred McClintic. Fred turned all that stuff into a computerized wargame called, with some justice, the *McClintic Theater Model* (*MTM*). Now over 10 years old, it is still one of the more popular models in the military. Ray liked to cast me as one of the "fathers" of *MTM*. But that's too generous; at best I was a godfather. It was Ray and Fred who pushed the project to completion, on time and at a fraction of the cost of similar military wargaming projects.

In 1981, in recognition of Ray's five years of effort, a wargaming department was formally reestablished at the Army War College. In 1984

I turned down an offer to become a "Professor of Wargaming" at the war college. I felt that it was much more effective, and safer, to give advice from the outside. The title did have a nice ring to it, though. This professorship was offered partially as a result of Ray Macedonia's retirement. The primary reason for Ray retiring was medical. The aftereffects of his tour in Vietnam were catching up with him, and it seemed a good idea to get out while there was still time to have any retirement at all. Ray's departure left the Wargaming Department to the vagaries of bureaucratic infighting, and it ceased to be a department for a while. But the point had been made, and wargaming survived. Wargaming had come full circle at the war college. First established there at the turn of the century to avoid planning and mobilization disasters in a future war, the war college was again being used to plan the army's move into the next century.

The curious aspect of all this is why the army moved away from wargaming in the first place. The army had abandoned wargames just before World War II because of a feud with the navy and because wargaming never really took hold (officially) within the army. There were always individual officers (and some troops) who wargamed, but without strong official backing, these efforts had little impact on combat capability.

The feud in question came about when the army and navy were attempting to develop joint war plans in the years immediately before World War II. These "Rainbow" plans called for closer coordination of army and navy efforts than ever before, and neither service was able to square its own strategy with that of the other. It took efforts by the president and several other senior officials to get (and keep) the army and navy working together. One result of this split was that the army did not get full use out of the extensive (and quite accurate) wargaming the navy had done during the 1930s on the subject of a future war with Japan.

After World War II, another attempt was made to get the army and navy to work together. One of the major areas of cooperation was to have been wargaming. Even though the army transferred most of its wargaming assets to the new National War College in Washington, D.C. (each service has its own war college; the army's is in Pennsylvania, the navy's in Rhode Island), the navy refused to cooperate, and got away with it this time. The National War College turned from the study of war to an emphasis on politics and management. The army never tried to retrieve its wargaming program, and wargaming disappeared from army schools for over 30 years. It also disappeared from the National War College.

In another one of those oddball ironies, the Army War College wargaming operation got another shot in the arm in the mid-1980s as a component for a new "school for generals." Odd as it may seem, new flag officers (generals and admirals) have not traditionally been given any specific training on how to operate in their new, and exalted, ranks. Flag officers are an entirely different breed in the military. While they don't walk on water, many of their subordinates are encouraged to believe the superbrass are capable of similar wonders. Fortunately, one of the earliest projects of the new Wargaming Department at the war college was a "theater level" wargame (*MTM*). Such a game would be the type a general (a senior one at that) would deal with. The lieutenant-colonel students at the war college were on the fast track to flag rank, and such a game was intended to give them a taste of what command would be like if they ever made flag rank (and about a third of them did).

All this wargaming that allowed lieutenant colonels to command armies generated the idea for a school for generals, once they became generals. The air force was also working on the idea of a "school for generals," and had dragged me in to work out some of the concepts. Initially, there was a lot of enthusiasm for a united (all services) school for flag officers. But this never really got off the ground, and eventually the army came back to the war college to get one going for its own newly minted generals. While it wasn't the only reason for keeping wargaming alive at the Army War College, it was a big help.

Amid all this enthusiasm for wargaming, it was decided in the early 1980s that a wargaming center was to be set up at the National Defense University (or NDU, an all-services military graduate school in Washington, D.C.). I was asked to set up this center and serve as its first director. While I declined this offer, I did assist in finding a suitable director, and the NDU center got going and continues to operate.

## THE PAYOFF, AND WARNINGS ✱✱

Wargaming has greatly influenced DoD decision making since the 1970s. While the older computer-driven models are still used to forecast future needs for logistics and personnel levels, the manual games are now created all over the place to support local decision making. My lectures at the Army War College and similar places stressed the relative ease with which one could develop useful wargames on the spot. Fifteen years after the first of these lectures in the mid-1970s, that advice was bearing fruit. Thousands of officers went to work on existing games and new designs. In 1978, for example, an army officer stationed in Thailand sent a game to the Army War College that was incorporated into the curriculum. There were similar cases closer to home.

This profusion of manual games based on the principles of historical wargame design, particularly the need for validation, had an increasing effect on all existing military wargames, models, and simulations. By the early 1980s, it became increasingly difficult to shrug off real or imagined criticisms of unrealistic results in the older models. This created a fair amount of ferment in the modeling community during the late 1980s and early 1990s. What it came down to was that many of the older models used to generate analysis to support DoD budgets and military planning were probably quite inaccurate. No one on the inside is eager to admit this publicly, as these are often the same people who have stood behind, and sometimes created, these models for several decades. The collapse of the Warsaw Pact will have a lot to do with avoiding any real embarrassment in this area. The old models, focused as they were on the potential war with the Warsaw Pact, are now being quietly retired and replaced with a new generation of models. If not replaced, the older models are having their analytic guts replaced with stuff that is more likely to conform to the real world.

This does not assure the reality of models in the future. While history is a good guide, indeed an iron guide, for historical wargames, there is more opportunity to rewrite the history of battles not yet fought. It's now widely known that previous models often were modified so the results would conform to what the current defense policy was. This can still happen. A case in point is the extraordinarily low casualties in the Gulf War. This was off the scale as far as historical battles go and, while it may indicate a trend, it's not going to be the norm anytime soon. But that doesn't mean that Congress won't demand that all future military actions be measured by how low the casualty count was in the Gulf, or how low some politicians thought it should be. The politicians also determine what the military objective is when they order the troops out, and even if the officers warn of high casualties, politicians will crucify any commander who loses more troops than the current political wisdom thinks should be lost. As a result of this mind-set, there will be pressure on model builders to show low casualties, or at least shave the numbers a bit downward. Future wars have not been fought, so you can make adjustments, and making adjustments is the norm, not the exception.

This trifling with the numbers is nothing new, and I have personal experience with it. My first run-in with this drill was in 1976. I was designing *Firefight* for the U.S. Army Infantry School. The people there wanted a realistic game of small-unit combat in a Central European environment. Part of that environment was a lot of underbrush and trees. We had people go out, literally, into similar terrain and check the lines of sight (unobstructed view) gunners would have. The game showed that you rarely got a clear shot for more than 500 meters. Several Bundeswehr studies of West German terrain in the 1960s and 1970s showed the same effect. When our Infantry School liaison saw the map and realized that tanks were not going to be able to use their long-range guns to full effect, he said something would have to be done. I suggested another game map showing desert terrain. No good, we were told, the emphasis must be on the wooded terrain of Central Europe. And because the army was still diligently studying the Israeli experience in the 1973 war, it wanted the troops to experience the usefulness of long-range tank fire. After much talking back and forth, it was agreed that we would strip the underbrush and many of the trees from the game map (which was taken from actual army maps of training areas in Germany and Georgia) and even flatten a few troublesome hills. Now the troops could get those 2,000-meter shots their commanders were so fond of. In the game, anyway. It's a good thing the Warsaw Pact collapsed in 1989, because there were still a lot of U.S. Army commanders who had visions of long-range tank shooting in Central Europe. After the incidents of 3,000–4,000 meter shots in the Iraqi desert, many probably still believe they'd get the same results in Europe. Fortunately, the troops (and I was one for three years) are remarkably free of the delusions of their commanders. Unfortunately, the commanders make the policy, give the orders, and generally structure the environment the troops will have to fight in.

Another example was the second change the Infantry School wanted to make in *Firefight*. As was my practice in tactical-level games, I made an allowance for "command control" problems. That is, you could be certain that some of your orders would not be carried out, and you never knew exactly which ones. Bad communications, inept subordinates, misunderstandings—the reasons are endless. Every war and every army suffers from these problems to one extent or another—the friendly-fire casualties in the Gulf War were just one of the more obvious manifestations of this problem. In the games, I assign a degree of command control difficulty to each unit in proportion to its historical performance. The Infantry School dismissed any problems with command control out of hand. "The U.S. Army doesn't have any problems with command control," I was told. Telling a former enlisted man in the combat arms such nonsense was a big mistake. I'd been waiting 12 years (since my discharge) to chew out an officer for assuming he was in complete control in the field. So I had a good time of it, explaining the battlefield facts of life to the poor wretch from the Infantry School. No go, however, so in *Firefight* command control would function as every proper commander wished it would. The army obviously was not yet ready for the lessons of history. Vietnam was at that point still in the category of a bad memory. In a few years, the Vietnam experience did become history, was studied, and by the early 1980s command-control problems were being addressed. You've got to have patience in this business.

Sometimes patience isn't enough, though, and events overtake poor modeling and wargaming. For nearly 50 years, the Warsaw Pact and NATO armies faced each other in Central Europe. Throughout the period, U.S. modeling always predicted a high risk of the Russians quickly sweeping to the Rhine, and beyond. In all the modeling I have done, this was rarely the result. When all the quantitative and qualitative factors of both sides were taken into account, the Russians were faced with a slender chance of victory, at best. Ironically, on the other side of the Iron Curtain it was taken as an article of faith that the Russian divisions would cut right through the NATO defenders. In 1989, *glasnost* allowed several Russian General Staff model builders to admit that their studies had also shown that Russian victory in Europe would be unlikely. But until recently it was considered a career-threatening move to push these history-based models too forcefully.

The next round of blowing smoke will involve restructuring the armed forces for the vastly different military needs of a post–Cold War world. The armed forces need accurate models to determine how well, or poorly, they can carry out the missions given them by the government. We were lucky in the Gulf, we finally got to fight that big air/armor battle anticipated in Central Europe. We were well trained and well equipped, and instead of the Russians, we faced the less capable Iraqis using Russian equipment. Fate may not be so benevolent the next time around.

But fate, in this case, has less to do with the effective use of wargaming than with the presence of individuals who understand wargaming and are willing to stick their necks out to initiate and preserve wargame use. Wargames are very complex beasts, and few people are capable of understanding them. This is unfortunate, but in my 30 years of working with wargames, I have been repeatedly reminded that the talent for understanding wargames is spread quite thinly throughout the population. Probably no more than a few percent of the population can grasp the internal concepts of wargames. Somewhat more people can use manual wargames, or at least go through the motions. Wargames running on computers are much easier to use, as there is less about the internal workings of the game you have to understand and more clear-cut decisions you can make. About a quarter of the population can profitably handle a good computer wargame.

Because of this, the history of wargame use in the military is replete with situations where wargames disappeared simply because there weren't enough people with the wargame "talent" available to keep wargames in use. For this reason, wargames disappeared from the U.S. Army before World War II, and didn't reappear until the 1970s. Even the U.S. Navy, which long ago made wargames part of its very fabric, has had periods where the wargame centers were largely run by people who could use wargames but didn't really understand what they were all about. The U.S. Air Force, created from the Army Air Force in the late 1940s when wargaming was largely forgotten in the Army, took over twenty-five years to rediscover wargames. This was an interesting development, as the highly technical air force had plenty of well educated, scientifically inclined people, a group that produces most wargamers. But until a potential wargamer discovers what wargames are, and then rises to high rank, no one will be issuing orders to begin using wargames. Eventually, the air force did this. But it took a quarter of a century.

Designing wargames requires a special, and rare, set of mental tools. Go through Chapter 4 of this book again if you have any doubts about this. The manual wargames, unlike the computer wargames, bring players into intimate contact with the details of the games' design and mechanics, and for this reason there are very few people who can handle playing the manual games. In effect, the audience for manual wargames is pretty much restricted to those who are capable of designing them. Early on, I realized this and found that it took only a little pressure to convince potential designers that they could do it, and many of them did. As a result, the several dozen designers I trained in the 1960s and 1970s can be found today designing a lot of the games used by the military (as well as doing commercial products, and sometimes both). Many more wargamers simply took me at my word: "If you can play them, you can design them." Computer wargames are a different matter, as you do not have to personally manipulate and comprehend all the details of such highly automated games. To design a decent computer wargame, it's still safest to go back to a manual design first.

The military is run in such a way that unique talents are not often recognized, much less put to use. Those with wargame-design talent rarely find themselves in jobs dealing *with* wargame design. Even then, most military people are only temporarily in an assignment before moving on. This is a major reason why civilian consulting firms are hired to do a lot of the wargame work. The civilian firms seek out people with specific talents and keep them in the same job for a long time. But for many decades, even the consulting firms could not recognize wargame-design talent. This situation created all those models, simulations, and wargames between 1945 and the 1970s that, by and large, were not of much use. Through most of this period, the U.S. military high command (the JCS, Joint Chiefs of Staff) maintained an organization devoted, in theory, to wargaming. Whether it was called the Joint Wargames Agency, SAGA, or whatever, it always had only people whose turn it was to serve with the JCS, and rarely were these people who really understood what a wargame was. Ray Macedonia was one of these, serving on the staff of SAGA for several years. Ray could understand a wargame, yet until the mid-1970s, he'd never seen a real one. Once he did, he went looking for other wargamers and got the show on the road.

In the past, similar individuals in the right position at the right time either discovered wargames or reinvented them. But these situations are rather rare, and for this reason, wargaming has always been a sometime thing. When I was first called down to lecture at the Army War College in the 1970s, I immediately realized that the most important thing to do was to make as many people there as possible aware of what a wargame was (whether they understood them or not) and to drive home the point that there were many wargamers in the military, and all you had to do was find them and put them to work.

Thus, for over 15 years, I have been making the same point about knowing what a wargame, and a wargamer, is, finding them, and letting these rare individuals perform their magic. This approach has worked. As long as it continues to work, wargaming will not disappear again.

One wargame area where the OR crowd did shine was in vehicle simulators. Over half a century ago, the first aircraft simulators were developed to aid pilot training. As computers became more powerful, so did the aircraft simulators. Today, they are remarkably similar to those you can buy to play on your PC, although the military ones often cost more than the aircraft they simulate and are extremely realistic. Simulators are also available for complex ground vehicles (tanks and APCs). Simulators, as the name implies, simulate the operation of a specific vehicle. This is a much less complex task than simulating an entire battle. It works, and as computers become more powerful, it also works with a larger number of simulators operating together. So the "OR" approach to wargaming has reached the point where OR can accurately re-create several aircraft, or several dozen armored vehicles, in combat.

One of the most ambitious ground vehicle combat simulators is Simnet. This is basically a network of inexpensive armored vehicle simulators built from off-the-shelf components. Moreover, the PCs that run these individual simulators are electronically linked to others in the same area and still more Simnet simulators around the world. Thus, a tank battalion in the United States can, via satellite link, fight it out with a tank battalion in Germany. Simnet also provides helicopter and aircraft cooperation. Simnet has been in the works since the late 1970s and has been quite popular with the troops, as well as very effective. The army has wisely decided to expand Simnet use and improve the technology. Simnet is a rare example of military procurement where civilian technology is quickly, and cheaply, adapted to military use.

The air force, for all its technology, never got into wargames in the traditional sense, but instead invested heavily in flight simulators. It had some planning models for logistics, and developed one of the first military computerized parts-inventory systems. But during the 1980s, the Air Force War College spent a lot of money setting up an Air Force Wargaming Center. For the first time in the nation's history, all three services (plus the marines) were heavily involved in wargaming.

## WARGAMING ABROAD ✱✱

The U.S. Army's abandonment of wargaming had enormous impact on foreign armies. As the U.S. Army was the most powerful ground force in the West, all of America's allies tended to follow its lead in, or rather away from, wargaming. Before World War II, Germany was the leader in wargaming developments, but after World War II, Germany no longer played a leading role in any aspect of military affairs. Thus, America's superpower status and rejection of warmaking after World War II put most wargaming out of action for 30 years.

Britain was quick to go over to the operations-research approach, although the British armed forces did maintain a keen interest in historical studies. Britain had been one of the pioneers in operations research, so that between the efforts of U.S. and British OR experts, the purely OR approach to wargaming took firm hold in Western military thinking. Most other nations also maintained a respect for historical military studies, but this counted for little as the United States generally led the way in new military developments.

Russia, on the other hand, still maintained a wargaming tradition, but one that was not exactly wargaming and, oddly enough, resembled the U.S.-inspired operations-research approach. As with much of their science and technology, the Russians took wargaming ideas from the West and turned them into something uniquely their own. Put simply, the Russians took the systematic German historical approach to the study of military affairs and the U.S. operations-research techniques, and combined them. What resulted were massive studies of past military actions and the application of operations-research techniques to these studies to produce predictions of what could be expected in future battles. The Russians' work is quite impressive. Their historical research is first-rate, and they have a firm grasp of operations-research techniques. Two other Russian habits, however, have hobbled their work. The first is the Russians' mania for secrecy. Few Russian researchers had access to the historical military research, or even the source data in the closely guarded archives. Second, like any other "theoretical" military work, the results of wargaming had to conform in some way to Marxist-Leninist dogma. This forced the Russian wargamers to recast their results into sometimes erroneous and misleading forms. The result was that, like most Russian science and technology, the theoretical basis of the Russians' wargaming was excellent, but secrecy and politics prevented them from doing much useful with it. Russian wargames are quite similar to the spreadsheet-based wargames discussed elsewhere in this book. They use operational and operational-strategic models (especially since the mid-1970s). They make extensive use of carefully verified and intensively analyzed historical models. The battles of Kalkhin Gol (1939) and Kursk (1943) are favorite campaigns used for study. Russian wargamers who have come up with conclusions that did not reflect "correct thinking" found their careers in danger and their work ignored. Ongoing reforms in Russia have changed this, and once Russia does manage to reform its military on a wide scale, its wargaming will emerge from the shadows and make a substantial contribution to our understanding of how warfare works, in the past as well as the future. There are already many changes occurring in Russian wargaming as a result of the fall of Communism in 1989. U.S. and Russian wargamers have been meeting and comparing notes since 1990, and we can expect some interesting developments out of that.

## WARGAMES AND THE 1991 IRAQ WAR ✱✱

Wargaming featured prominently in U.S. efforts during the 1990–91 war in the Persian Gulf. On the morning of August 2, with Iraq's conquest of Kuwait still not complete, the Pentagon looked around for some quick wargaming on what was going on and what it all meant. The only kind of wargame that could get results quickly was a manual game, a commercial manual game that could be bought in a game store. The game used was *Gulf Strike*. Mark Herman had designed this game on potential wars in the Persian Gulf during the mid-1980s. The game had already been updated once a few years later and was still in print. Mark had, for several years, been working for one defense consulting firm or another, so the Pentagon knew who he was and what he could do. The Pentagon approached Mark at 10:00 A.M. on August 2, he was under contract at 2:00 P.M., and the game began at 3:00 P.M. (using various Pentagon Middle East experts as players). Before the day was out, Iraq had conquered Kuwait, but the wargamers in Washington knew Iraq was doomed. The results of this manual game were the basis of most of the decision making during August. Ironically, when Mark went to update *Gulf Strike* for commercial release, he had to borrow the Order of Battle (and other) information from the *Arabian Nightmares* game Austin Bay and I were working on for late 1990 publication. Mark had used classified information for the August 2 *Gulf*

*Strike* game, so he had to get a new set of clearly unclassified information for a commercial version of the *Gulf Strike* update. Since Mark was another member of the old SPI gang, we had no problem giving him permission to use all he wanted from *Arabian Nightmares*. Many of the numerous SPI "school of wargame design" graduates were prominent as analysts on TV during the war. One wargamer quipped that "every time I turned on the TV news, it looked like an SPI reunion." For many years, SPI was the premier place to learn how to design wargames, particularly games on contemporary subjects. After leaving SPI, many of these designers went to work for the military or intelligence agencies.

Naturally, most of the officers and many of the troops in Desert Storm had used the dozens of different wargames developed during the 1980s. Before Iraq invaded Kuwait, wargames were used to determine what kind of force would go to the Gulf if there was a war. There had been contingency planning for a Persian Gulf war since the early 1970s, and the planning had got even more intensive after Iraq invaded Iran in 1980. This pre–August 1990 wargaming did not deal so much with the nature of the future combat as with all the behind-the-scenes issues that had to be settled first. Among those issues were calculations on the size of the logistical effort required to get the troops there and sustain them. The composition of the forces sent had to be worked out, as well as plans for the use of air power. Professional wargaming gets involved with a lot of dreary details that commercial wargames do not treat in depth. However, in war victory goes to the side that is best able to cope with the details.

The operations of the combat units on the battlefield were gamed out, using games similar to those available commercially. As the Persian Gulf was a desert zone, it was expected that the fighting would make extensive use of mechanized forces. This was the war the U.S. Army had been preparing to fight in Central Europe against Russian armies since the 1950s. It was always assumed that the same tactics and weapons would do in the Gulf. One thing the combat wargaming also had to do was calculate expected casulaties. This was a touchy subject, and always had been. It's also something of an unexamined area in the Pentagon. In 1988 I was invited down to the Department of Defense Medical School to give a series of lectures. One of them was for the faculty, and some officers from the Pentagon, on calculating losses in combat. In the course of my talk, I turned to the Pentagon group (whose specialty was calculating losses in future wars), and asked how they did it. The reply was, "That's why we came over here." They weren't kidding. Nothing much had changed by 1990. When the wargaming group of CENTCOM (the Central Command Headquarters controlling all coalition forces in the Gulf) was asked to come up with some casualty figures for a briefing back in Washington, they basically referred to average losses per day of combat for battles going back to World War II. They didn't need a computer or wargame for that. But when they were asked how many Iraqis would have to be killed or wounded before an August 1990 advance into Saudi Arabia would be stopped, they came up with "50 percent." I had an opportunity to ask some of the key people involved where that number came from and was told, "The *TACWAR* game." *TACWAR* is one of the many computerized wargames currently used by the U.S. Army. So I turned to one of the folks (no names, to protect the innocent) responsible for building *TACWAR* and asked how the 50 percent figure was calculated. After being led around the mulberry bush a few times, I discovered that "someone had picked up some numbers somewhere," and it was these attrition and "unit ineffectiveness" formulas from "somewhere" that created the yardstick of "destroying 50 percent of Iraqi forces to render them ineffective." This 50 percent figure was used repeatedly. Now you know where it came from. From somewhere. At least it worked.

The wargame operation assigned to CENTCOM was small. Normally, CENTCOM is located at MacDill Air Force Base (outside Tampa, Florida). CENTCOM is the American military fire brigade for any emergency situations in the Persian Gulf area. There are similar headquarters for all major U.S. area commands (one for Europe, one for the Pacific, etc.). The CENTCOM wargame operation had about two dozen people, half military and half civilian (usually consultants, and mostly programmers). They had lots of hardware, a VAX 8650 and a VAX 4000 minicomputer. Also two Sun network servers connected to 22 Sun workstations (industrial-strength PCs). Numerous PCs were available in Saudi Arabia. They had three wargames available:

• ***TACWAR*** (division level). This did most of the work. Details of how it works further on in this section.

• ***TAM*** (an operational-scale wargame designed by Mark Herman, derived from the *SAS* game Mark and I did at SPI). *TAM* exists in both manual and computerized formats. Although CENTCOM didn't use *TAM*, it was used a lot in Washington, and using *TAM* in September showed that by October, Saudi Arabia would be safe from anything Iraq could throw at it. *TAM* also showed that a lot more troops would be needed to force Iraq out of Kuwait, and even more to keep the U.S. casualties down.

• ***JTLS*** (an updated *MTM*, the game I helped develop at the Army War College). Because this *JTLS* update of *MTM* prevented users from playing with the procedures within the game, it was not a useful tool for an ongoing war.

*TACWAR* was the game used the most, because it was able to deal with the most detail (logistics, in particular). Game turns were one or more days, after which the user changed decisions and objectives and ran the next day or more than one day. Each day of simulated operations took one to three minutes of real time.

You could run a large number of days all at once, getting the time per day down to less than a minute. Speed of execution, of course, depended on the speed of the computer you ran it on. Normally, it runs on a VAX minicomputer, although in theory, any minicomputer or 32-bit PC running Unix (a computer operating system) could handle the game.

When the war began, part of the CENTCOM wargame operation was in one of the first cargo aircraft heading to Saudi Arabia. The part that went to the Gulf had nine troops, the VAX 4000 (a bit larger than a PC and more powerful than the older VAX 8650), one of the Sun network servers and seven of the Sun workstations. While *TACWAR* did a lot of the number-crunching to calculate things like logistic needs and the movement of major units into the area, the workstations were used for spreadsheet modeling, report writing, and preparing graphics for briefings. *TACWAR*, running on the VAX, took 15–20 minutes for each simulation of 30 days of operations. In most respects, *TACWAR* was a traditional military wargame. That is, you put in your assumptions about who could do what to whom and with what, and *TACWAR* would perform a lot of calculations. *TACWAR* did not allow for players, as such. The user loaded in what he thought each side had and what it would do. The "Artificial Intelligence" routines that caused each side to make this or that decision during combat were not as sophisticated as those found in commercial wargames. But *TACWAR* was able to do a credible job nevertheless. Sort of like a commercial computer wargame where you let the computer play each side and just wait for the results. In the case of *TACWAR*, the results are in the form of detailed reports and pay a lot more attention to logistics and support. *TACWAR* could deal with air, ground, logistics, and chemical-warfare operations as well as new units arriving during the period covered by the game. *TACWAR* did not have any fancy output (although that is in the works). You got page after page of numbers and cryptic terms. A freelance PC-based graphics package was used to pretty up the *TACWAR* results before passing them on to CENTCOM staff and commanders.

Manual and spreadsheet modeling were used to do a lot of the logistics work and sorting out how best to use the air power. In other words, a lot of the wargaming was done with tools available and familiar to civilian wargamers. Civilian wargamers are generally shielded from most of the logistic and support details that are needed to make an armed force functional, but the military must pay close attention to these items. It was the computerized logistical exercises that made it possible to launch the ground attack around the Iraqi flank. Considering the millions of tons of supplies (and each item getting to the right place at the right time), over 100,000 vehicles and half a million troops involved, you can see how only a computerized wargame could handle such a load. *TACWAR* simulations indicated the best places for supply dumps, how much tonnage could be sent up which roads (or cross-country), and if enough supply could be moved in time to support certain types of operations. A lot of credit for the apparent smoothness of Desert Storm has to go to the CENTCOM wargamers.

I don't know how many of the CENTCOM crew went through one of my lectures, but they certainly were using every tool in my toolbox. Order of Battle data was kept in a commercial data-base program and dumped into *JTLS* and spreadsheets for analysis. Manual wargames were used, and modified as needed. Of the two computer wargames available, *JTLS* was used only to generate tactical deployment on the computer screens. These screens were then turned into slides and overlays for the many briefings that had to be given. Briefings were very important; the best wargaming in the world is worth little if you can't present the results quickly and clearly to the commanders who have to make decisions.

While a complete wargame operation was in Saudi Arabia, there was another one (the original equipment and troops not sent to the Gulf) back at CENTCOM Headquarters in Florida. The two wargame operations had access to the same data and wargames. Data was transferred quickly through a satellite link between Florida and the Gulf.

As this was the first time U.S. wargames went to war for all services, there were some problems. A major problem was keeping *TACWAR* up to date, as the game was still a pretty large piece of computer programming. While commercial wargames spend a lot of time making their products "bulletproof" (unlikely to fail when operating), military computer wargames assume there will always be programmers around to tinker with the program as needed. *TACWAR* had programmers available for on-the-spot enhancements and bug fixes. But the programmers did not have security clearances high enough to allow them to work on *TACWAR* while it was loaded with the latest data about coalition troops in the Gulf (much less future plans). So when the programmers had to work on the system, a different (unclassified) set of data had to be loaded. The security problems went beyond programmer access. There was very tight security in Saudi Arabia concerning planning data. In addition to the usual keeping things locked up, troops were only allowed to know secret data that was required for their particular job. For the wargamers, this was a constant problem as they had to know everything in order to wargame out all the options the senior commanders were playing. As this was the first American war where wargaming was an integral and ongoing part of the command process, most of the people involved were not accustomed to sharing everything they had with a bunch of relatively low-rank people (mostly majors) mumbling something about games. Several times, it required the intervention of the CINC (commander in chief of CENTCOM, General Schwarzkopf) to get the data flowing to the wargamers. This tight security caused problems for other groups also, but none were hassled by it as much as the wargamers. The wargamers had to know everything in order to wargame out what hadn't happened yet.

General Schwarzkopf went through the Army War College a few years before I began to give my lectures there. However, he had got religion as far as how much a commander should use wargames. The CENTCOM wargames crew was nearly worked to death, with overnighters common. When the CINC's staff dropped a request in their laps, they had from six to thirty-six hours to get an answer. The actual gaming didn't take that long, but gathering the information from all the units involved did. There was no centralized reporting for what every unit in the area was doing, could be doing, or planned on doing. Wargamers required much more information than anyone else in the commander's staff, and this was not fully appreciated until the commander began to rely on the wargamers to constantly check out his ever-changing options. There were often several new scenarios to check out each week, in addition to updating the existing data base of units and material (supplies) on hand. Granted, the wargame crew worked in an air-conditioned bunker. But they were some of the hardest-working troops over there. And during many Scud alerts, the air-conditioning was turned off (to prevent poison gas from being drawn inside).

The Gulf wargamers were constantly called upon to give an update on expected casualties. Casualties were a hot political item back in Washington. The troops in the Gulf were also concerned about casualties, but they weren't worrying about elections. One of the first *TACWAR* wargames in August of 1990 set the then-minuscule coalition forces against 23 Iraqi divisions trying to sweep down the Gulf coast toward the major Saudi ports and airports. This game showed 20,000–25,000 coalition casualties. But as the coalition forces poured into the Gulf, the casualty numbers coming out of *TACWAR* went down to about 2,000 by early February 1991. The final estimate was higher than the actual losses, but then *TACWAR* was able to make the Iraqis fight back, while Saddam couldn't.

As the first heavy wartime use of wargaming, a lot was learned about what had to be changed, improved, and added. It's a long list:

**1.** Wargaming has to be at the center of everything. If you want your wargames accurate, they have to reflect the reality of your current situation as much as possible. In a pinch, you can make estimates. But the more accurate the data in the wargame is, the more accurate advice the wargame will give you.

**2.** Wargaming has to be part of staff operations. Historically, the commander's staff are the information gatherers and analysts who provide expert opinions on what will work best. The commander then decides what to do. A lot of staff officers are still a little leery of wargames. Some feel threatened, afraid that some (or all) of their jobs will be replaced by a wargame (with or without a computer). Some of these staff officers are correct, but they are generally the glib deadwood you could do without anyway.

**3.** Keep everyone happy. There are dozens of different "unions" the wargame has to serve, and serve well, to be effective and convincing (or is it the other way around?). These "unions" comprise the different military specialties: infantry, armor, artillery, helicopters, air-force fighters, bombers, tankers and transports, various navy specialties, supply units, maintenance, medical, engineers, special forces, and so on. All have to see convincing results from the wargame for *their* particular activities. Military wargames have always tried to do this, and often did it at the expense of the combat arms, as until Desert Shield/Storm, wargames didn't go to the front. Now they do, and to survive, wargames must please everyone.

**4.** Wargame faster. The primary reason CENTCOM bothered to take the wargames crew with them was because they knew they could get fast results. This was only possible with the new wargames developed during the 1980s. As wargames become even faster, they will be relied on even more. By the end of the decade, commanders will be running the show from their keyboards, monitoring the situation on a CRT.

**5.** Clear output. Cryptic output was a major shortcoming of *TACWAR*, and a lot of time was spent translating that output into something the commander and staff could comprehend. Commercial computer-wargame designers know all about this problem, and I'm pretty certain that a lot of their solutions will be borrowed by the military for the next generation of professional wargames.

**6.** Build confidence. Failure travels farther and faster than success. It has taken most of the 1980s for the wargames community to generate enough successful work to win the confidence of some of the senior commanders. If the CENTCOM commander and staff were not confident in their wargame staff's ability, they would not have taken them to the Gulf. Wargame operations in the Gulf were generally successful, but many users came away aware of shortcomings. Military wargamers will have to build on their success and address their shortcomings in order to keep the confidence in wargaming high. There are still a lot of senior officers who have a dim view of wargaming's worth on the battlefield.

**7.** Be neutral, objective, and convincing. Wargamers have a lot of power if they have any credibility. The Gulf experiences showed how commanders and staffs would use wargame results to settle disagreements over how to proceed. But since the uniformed wargamers are drawn from a large range of other specialties, they all have (or are suspected of having) stronger loyalties to those original areas. Wargaming is not yet a separate specialty, and officers from any one of hundreds of military job specialties are trained in the "secondary" specialty of wargaming (the colonel running the wargame unit formerly flew C-5A transports for a living). Eventually, they will go back to their primary job, and there is always the suspicion that their wargaming advice will color their recommendations. So far, such favoritism has not occurred to any large extent. It was not an issue in the Gulf. But you always have to be careful.

**8.** More secure, more reliable, and more flexible wargames. More speed and better output will not be enough for the next generation of wargames. *TACWAR* was the first of the new generation of wargames, or the last of the old generation, depending on how you look at it. It has got faster simply by running it on ever-faster computers. Better output can be added to the existing program. But experience in the Gulf has demonstrated that this will not be enough. The security people were not happy with how accessible highly confidential information was on the *TACWAR* computer. A combination of improvements in software and hardware will fix this. Reliability was another problem. Military wargames are not as stable as commercial products. Partially it's a tradition, of not enough people looking for problems and not enough quality control. That can be fixed, if old habits can be broken. Flexibility will come as a result of the Gulf experience. Now wargamers know what they will be called to do in a combat situation.

**9.** Think big. Wargamers have to think like their commander, and look at the big picture. This has never been a problem with civilian wargamers; being in charge of their cardboard or electronic army has always been one of the primary appeals of wargames. But military wargamers are first, and often finally, staff officers. They have to stop thinking like clerks and more like commanders in order to get the most out of their games, or at least get what their commander needs.

The *TACWAR* crew was not the only wargaming operation involved in the Gulf War. Two months before Iraq invaded, CENTCOM conducted a large computer wargame postulating that Iran would invade Iraq and that the United States would side with Iraq. What was notable about this operation was that it used several different wargames and involved air, land, and naval forces. In July of 1990, the U.S. Air Force did a wargame of Iraq invading Saudi Arabia. Once the shooting began, the U.S. Marine Corps conducted a series of six wargames (the first two manual, the others computer-assisted) on possible future operations in the Gulf. A strategic game was conducted in August, concentrating on getting forces to the Gulf. In October there was a campaign game covering overall operations in Saudi Arabia and Kuwait. In November there was an operational-level game concentrating on marine forces. In December there was a breaching game, to work out how Iraqi fortifications could best be breached. Another campaign game was run in early February that predicted under 3,000 coalition casualties (on the assumption that most Iraqis would fight). In March there was a "War Termination" game, looking at what could happen after the cease-fire. The results of all these games were added to what the *TACWAR* gang was coming up with, and the results generally matched. The U.S. Air Force regularly ran simulations of its air operations, both before and during the air war.

Back in the United States, various agencies were also running manual and computerized wargames. Most of the military wargamers had used commercial wargames, and several stated publicly that they were influenced by them. The Iraqis, it turned out, were also quite keen on wargames, all of them manual. Some were of the miniatures type, using detailed terrain models. Iraqi wargamers were willing and able, though most Iraqi combat troops weren't.

As a footnote to all this, there was quite a lot of commercial wargaming activity on the Gulf War during the "waiting period" between the Iraqi invasion in August 1990 and the coalition counteroffensive in January 1991. In August of 1990, I was again editing *Strategy and Tactics* magazine. Some wargamers, aware of my past efforts in doing games on wars about to be fought, suggested that it was time to do it again. I didn't have the time to do it myself, but thought it was a neat idea and suggested that a friend of mine with wargaming experience do it. Austin Bay was the fellow in question, and he rose to the challenge. The design was complete in less than a month and appeared in print before the end of the year. The game, *Arabian Nightmares*, was right on target. It wasn't the only one. Mark Herman quickly came up with an "update kit" for his *Gulf Strike* game, got it into print by the end of the year, and was also accurate in predicting the course of the war. Several other games also came out after the war, treating Desert Storm as another historical event to be wargamed.

## CREATING WARGAMES FOR THE TROOPS ✱✱

The military doesn't design wargames the same way commercial wargames are put together. There are a host of special situations and problems it must contend with. Since the late 1970s, I have been called upon to give lectures, lasting from half an hour to several days, on how I feel it should be done. These lectures are quite popular, and I get invited back to some venues year after year. More important, I constantly run into military wargamers who have been successfully using the guidelines presented in these lectures. What follows is the advice I have been giving to military wargames designers over the last 15 years. This material has always been given in the form of a lecture, so it's about time to get it all into print. There are ideas here that even the designer (or player) of commercial wargames will find useful. There's no better way to understand the differences between military and commercial wargames than to compare what follows with Chapter 4's advice on designing commercial wargames. There are some interesting differences.

Some of the lectures last half an hour, some go on for several days. What follows is a recapitulation of all the items I try to cover. When I have more time, I go into more detail. Otherwise, I present a checklist format.

### THE GOLDEN RULES

All situations can be easily modeled using a half-dozen design rules and past experience with similar situations. The rules are:

**1.** *Know what the user wants.* It's difficult enough knowing what you want to do when you are doing a model for yourself. It's easy to start building a model with a vague idea of what you want. It's impossible to complete an adequate model unless you have developed a precise idea of what you want it to do. If the user is someone else, you have to help him figure out what he wants it to do. This is not easy, and is often avoided because of the difficulty. Don't avoid it; be difficult if you have to. In the long run, this is the easy way out. To define the needs of the project, apply this checklist. It will get you started in defining the model user's needs. If you can't define your project adequately, you'll waste a lot of time and effort. You probably won't complete your project either. The last thing you want to hear from the user is, "That's what I asked for, but it's not what I want."

**A.** *Determine the process to be modeled.* Many different aspects of your model must be defined before you can proceed. Scale (Strategic, Operational, Tactical), Environment (Land, Air, Naval, Combined),

Intensity (Low, Medium, High), Basic Aspects (Movement, Combat, Order of Battle), Special Aspects ($C^3I$, Logistics, Doctrine and Tactics, Fog of War—is the situation highly dependent on one or both sides being in the dark about what is going on? If so, you will have to model this aspect of the situation.).

**B.** *What do you want it to do?* There are several different tasks you can direct your modeling toward. These can include training, research, analysis, etc. For example:

*Test a hypothesis.* This can be historical, contemporary, or future. It can be about weapons, tactics, organization, or whatever. Be rigorous in defining your hypothesis. A model will eat you alive if you are sloppy.

*Define a process.* You may want to break down an existing system into its essential parts. A model-building exercise is excellent for this.

*Provide training.* There is no better way, other than actually going into the field with the system.

**2.** *Start with an existing model.* For example, to create a wargame for contemporary ground-combat operations, you can wander off to your local game or software store and see what the commercial designers are up to. There are also companies that deal in out-of-print games that may be of use. If there are any gamers in your area, buy them a few beers and pump them shamelessly for leads. There's also a lot of previous work in the noncommercial sector waiting to be plundered. No sense reinventing the wheel, especially since that approach is sure to lead to exceeding your budget and missing deadlines. Don't endanger your career. Plagiarize. There's no copyright on ideas, and most of the ones you need have already been thought of and thought out by more experienced designers. I know—I often steal from myself (as well as others—that's why I'm an expert).

**3.** *Be sure you know what you know.* Pick a subject you have a keen interest in, or have gained a perceptive knowledge of. This will eliminate a lot of time-consuming research. You wouldn't be doing this if you weren't an expert in something.

**4.** *Compile information.* Once you have agreed upon what you want to do, you must gather information. Here is a sample checklist:

**A.** *Area of Operations.* Where, in time and geography, is the conflict to take place?

**B.** *Scale.* What is to be represented on the map, a few square miles or a continent?

**C.** *Significant Terrain.* For the Terrain Effects Chart, this is a winnowing process, in which you reduce all the terrain information you have gathered into a usable format.

**D.** *Order of Battle.* Units involved, their movement capability, combat capability, and other characteristics.
**E.** *Victory Conditions.* This is a critical element, and often slighted or overlooked. What are the goals of the combatants?
**F.** *Combat Results.* Attrition rates in combat, with adjustments for other factors as needed and likely distribution of results for use with nondeterministic (unpredictability of combat) procedures.
**G.** *Sequence of Play.* Sequence that appears to work best in most situations is: 1. Planning and preparation operations; 2. Movement; 3. Combat; 4. Postoperations checks (victory, morale, command control, etc.).

**5.** *Integration.* The Big Moment—you create the prototype. This is where you assemble the first working version of the game. The prototype is usually quick-and-dirty. Just get it working, quickly. Once that is done, Check the switches. Whether the game is manual or computerized, you should have Probability Tables that can be easily changed to adjust the games outcomes in a controllable fashion. Finally, a note on "Pre-dawn Madness and the Bleeding Edge of Technology." There is a bit of magic involved at this point. The model must be exercised, errors noted, and the model modified and exercised again. Strange things will happen, and you will often find yourself spending more hours working on this phase than you realize. This is the Pre-dawn Madness most programmers are familiar with. Don't expect to understand everything that's going on in the prototype. If it works, leave it be and go on to the next item. Don't be any more inventive than you have to be. Beware the Bleeding Edge of Technology: Stay with the simple and don't get cute.

**6.** *Testing and User Acceptance.* First there is alpha-testing, where first you and then some typical users must be able to reproduce a historical event, or a defined hypothetical event. Then comes blind- (or beta-) testing, where the game is handed to typical users without you hovering over them ("blind" to you). Last, there is ongoing testing after installation. No model is ever truly finished.

## DIFFERENCES BETWEEN HOBBYISTS AND PROFESSIONALS ✱✱

Although hobby and professional gamers share many of the same techniques (and often the same games), there are some major differences between the two groups, differences that explain a lot of the differences in attitudes and accomplishments of the two groups. In short, these differences are as follows:

• Professional gamers are, well, professional. They get paid for it. To many professional gamers, it's just a job. For hobby gamers, it's an avocation. While there are many enthusiastic professional wargamers, all hobby wargamers are very much into what they are doing.

• Professional gamers cannot talk freely about what they are doing. Most of the professional wargaming work is severely restricted in terms of who can talk about it and where. Hobby gamers speak freely about their games, and this torrent of comment and criticism makes the hobby-oriented games much better for it.

• Professional gamers do not worship validation (being sure their games represent reality as much as possible). Most hobby games are historical games that, in order to work, must be capable of re-creating the historical event they are based on. This ability to re-create the historical event is also called "validation." Hobby gamers take it as a given that if a game cannot be validated, it's not worth bothering with. Nearly all professional games are on wars not yet fought, so validation in the classic sense becomes moot. However, there is a tendency for professional gamers (or at least their masters) to make up their "future history" as they go along.

• Professional gamers serve many masters, while hobby gamers serve only one (themselves). Because professional gamers are getting paid for it, they have to be responsive to whoever is paying them. Often this involves not just one boss but an array of officials. All of these bosses want something from the professional games, and often these demands are contradictory.

• Professional and hobby wargamers have somewhat different backgrounds. Until the 1980s, most of the professional wargamers had a computer and/or operations-research background. Hobby gamers have a strong interest in history and technical subjects (science, engineering, medicine, law, etc., including OR and computers).

• Professional and hobby gamers have different experience with games and simulations. Hobby gamers nearly all have experience with general board games (especially chess, plus classics like Monopoly, Risk, etc.). Naturally, the hobby gamers are familiar with commercial manual wargames and, increasingly, commercial computer wargames. Hobby gamers are rarely familiar with noncommercial ("professional") wargames, and professional wargamers are usually familiar with little else (except some of the general board games). Programming experience is much more common among professional gamers, as most of their games are still run on computers.

• Military experience is quite common among hobby gamers. The commercial games are more accessible than the professional ones, there are no security issues to worry about, and this allows military people to openly address issues that concern them. Civilians with military experience are also more prone to use commercial games. In a tradition that is now over 30 years old, military people and civilians use the commercial games to obtain a greater depth of knowledge on military affairs.

• Use of wargames. The major difference between hobby and professional wargamers is the way they use the games. Hobbyists are interested in experiencing history; professionals are more intent on doing heavy-duty analysis (thus the predominance of computers) and, increasingly, training.

Gamers also tend to be exceptionally well represented in a handful of professions. This says a lot about the nature of wargames, wargamers, and how the wargames work. Programmers, or people comfortable with this uniquely 20th-century exercise in logic and technology, are well represented in wargaming circles. Though many wargames now run on computers, the ones that still attract programmers are the manual games, where the player can still tinker with the logic and procedure of the game. Most computer wargames do not allow such access.

Since the introduction of personal computers in the late 1970s, an increasing number of wargamers have got into programming in one form or another. All of these are relevant to wargames. The most common form of programming a lot of people are exposed to is personal-computer spreadsheet (Lotus 123, Excel, Quattro, etc.). All of these programs feature a "macro language," which is, in effect, a kind of computer-programming language. Since most personal computers come equipped with the easy-to-use BASIC programming language, millions of computer users have learned to use it. These millions of recreational and occasional programmers are added to over a million professional programmers to create a ready market for game "simulations" of all kinds.

Military experience has had an influence on how hobbyists and professional wargamers approach their work. Increasingly, though, people without combat (or even military) experience work on wargames of all types. While much of the research needed to create a game requires more scholarly training than time in the trenches, there is a certain insight required that can only be obtained from being in the ranks.

Designers of commercial games have the historical record, and if they lack personal insight on how the military operates because they've never been there, they can just work a little harder until they figure it all out. Professional wargamers have a different problem. Their games are on future wars and, as such, they have no historical hindsight to keep them straight. The military tries to overcome this problem by getting the troops involved. Decades of officers playing commercial wargames have created a pool of wargames-savvy men to put to work on the professional games.

Another problem unique to the professional gamer involves the buying and selling of wargame material. Many professional wargames are still produced by civilian firms that in turn sell them to other civilian managers of military wargaming agencies. Often this is a case of the blind selling to the blind, with neither end of the transaction having a firm grasp of the subject.

## TYPES OF PROFESSIONAL WARGAMES ✱✱

While commercial wagames fall into only two types (manual map-based exercises and computer versions of the same), there is a far wider variety at the professional level. Models and simulations, both quite different from commercial wargames, are major elements of professional wargaming. Professional Combat Results Tables, for example, are called "attrition models," and the more elaborate are indeed models in a very real sense, repeatable and static representations of combat process. Professional wargames usually contain many models linked together in a system. This, of course, is the classic description of a simulation, along with its ability to handle multiple scenarios in a more interactive manner.

The primary difference between a wargame and a simulation is fuzzy, based on the concept that a wargame is not capable of multiple runs from which statistically significant results can be derived. For many manual games, this is true for practical reasons. However, once a manual wargame has been turned into a computerized version, you can let it play itself a sufficient number of times to obtain statistically significant results. The advantage of the manual wargame is that human players can obtain broader insights from it and become better able to deal with the intangibles of a situation. That said, some of the wargames described below which I've listed for comparison, are simulations, and all contain models.

**Manual Model with Map**— What is normally thought of as a "commercial manual wargame." (The original military wargames were of this type. But that was before computers and beltway-bandit consultants.) These have the following characteristics:

**1.** *Forces.* The Order of Battle, and all units involved in the simulation. This must be consistent with scale of model, and the optimum for manual playability is no more than 20 units per side.

**2.** *Movement.* Each unit is assigned a numerical value representing its ability to move across terrain.

**3.** *Combat.* Each unit is assigned a numerical value representing its combat ability.

**4.** *Map Display.* A paper map, optimally about 20 by 24 inches, or the distance that players can reach units without stretching. A hex grid is used to regulate movement and combat. Each hex cell contains a discrete type of terrain, which shows up on a Terrain Effects Chart listing its effects on movement and combat.

**5.** *Rules of Use.* Explicitly written-out procedures to operate the model. This also gives insight into the underlying process that drove the situation being modeled.

**6.** *Easiest Model to Create.* Best preparation is simply extensive playing of existing games.

**7.** *Inexpensive.* Paper is most common raw material.

**8.** *Paper Computer.* System organizes processing of information in much the same way as a computer, only much more slowly.

**9.** *Easy to Maintain.* Procedures are largely self-documenting because they are made obvious to the player. Otherwise, the model would be unplayable.

**10.** *Labor-Intensive to Use.* An average size game will take two to four hours to play to a decision. Larger ones take much longer.

**11.** *Not Highly Iterative.* Time required for each game takes too long. Replaying individual turns has value, and proceeds much more quickly. Numerous iterations are required of a game in order for its results to have statistical significance.

**12.** *Precursor of Computerized Version.* Programmer needs a manual model to work from.

**13.** *Time Required to Create.* 500–2,000 hours. Varies considerably with skill of creators. My personal record for a published model is 12 hours from cold start to tested prototype (*Battle for Germany* in one session from 6:00 P.M. to dawn, 1975). Another 100 hours required for testing and finishing rules. Lack of sufficient skill will make successful design impossible no matter how much time is spent.

**Manual Model Without Map—** Can be described as either a ''commercial manual wargame without a map,'' or as a staff study with easily modified parameters.

**1.** *Forces.* Same as model with map.

**2.** *Movement.* Deduced as a result of force on force calculations.

**3.** *Combat.* Adjusted force ratios compared and results computed.

**4.** *No Map Display.* Spatial positioning is derived from player decisions and results of combat. Somewhat abstracted.

**5.** *Rules of Use.* Same as model with map.

**6.** *Easiest Model to Create.* About as easy as the one with a map.

**7.** *Paper Computer.*

**8.** *Inexpensive.*
**9.** *Easy to Maintain.*
**10.** *Labor-Intensive to Use.*
**11.** *Not Highly Iterative.*
**12.** *Precursor of Computerized Version.*
**13.** *Time Required to Create.* 400–2,000 hours. Can be less time than map version because map does not have to be developed. Can be more difficult for the same reason if the model is complex.

**Spreadsheet Combat Model—** A "manual model without map" put up on a spreadsheet program.

**1.** *Similar to Manual Model Without Map.*
**2.** *Forces.* A larger Order of Battle can be handled because the computer keeps track of details rather than the player.
**3.** *Combat.* More complex combat routines can be used, again because of the computer.
**4.** *No Map Display.* One can be used off-line for reference purposes.
**5.** *Rules of Use Built In.* Allows users to be trained much more quickly.
**6.** *Some Graphics Capability (Charts).* Depends on the spreadsheet you use. Most have a graphics capability.
**7.** *Highly Iterative.* Lotus spreadsheet products have a data-table feature that makes it much easier to do sensitivity analysis.
**8.** *Constructed on Spreadsheet Program.* Recommended ones are Lotus 123, Quattro, and Excel. Improv also good.
**9.** *Time Required to Create.* 300–1,200 hours.

**Cost/Benefit Model—** A spreadsheet combat model optimized for evaluating individual weapons (or other item) performance.

**1.** *Similar to Spreadsheet Combat Model.*
**2.** *Measures Effectiveness of Weapons Systems.*
**3.** *Forces.* Extensive list of weapons systems and variations can be analyzed.
**4.** *Combat.* System on system, duel-type engagements. Synergism of many systems must be abstracted.
**5.** *No Map Display.*
**6.** *Some Graphics Capability (Charts).* Important for this type of analysis, as a large amount of numerical data is processed, and graphics makes all this easier to absorb.
**7.** *Rules of Use Built In.* The "rules" are the formulas for processing the numbers.

**8.** *Constructed on Spreadsheet Program.* Lotus 123, Quattro, or Excel are best for this because of their full array of spreadsheet commands. They also have windowing and easy-to-use, powerful command languages. Any other spreadsheet will handle the essentials. Lotus 123 is still especially good for these jobs, particularly because of the large number of add-on programs available for handling linear programming, data-base file access, and other tools.

**9.** *Highly Iterative.*

**10.** *Time Required to Create.* 300–1500 hours. Depends on how elaborate you want to get.

**Expert System Combat Model—** Models of expert knowledge on various aspects of combat operations.

**1.** *Forces.* Generally not modeled as extensively as in other simulations.

**2.** *Movement.* Abstracted.

**3.** *Combat.* Handled in richer detail, usually in form of interrogation.

**4.** *Normally No Map Display.* Not needed in most cases, but can be added.

**5.** *Rules of Use Built In.* Especially decision-making rules and options.

**6.** *Highly Iterative.*

**7.** *Heuristic.* Depends on how you set it up.

**8.** *Natural Language Interface.* Most of the Expert System shells allow user to communicate with system in plain English.

**9.** *Created on an Expert Systems Generator.*

**10.** *Can Be Made Part of a Computer Wargame.* Works best as complement to other forms of combat models.

**11.** *Can Be Written from Scratch in LISP or PROLOG.* This takes much longer than using a shell, although the shell version can be done first as a form of prototype. This will make it cheaper to do in LISP or PROLOG, as the systems analysis and design will be much more complete.

**12.** *Time Required to Create.* 200–2,500 hours. Depends on several variables. Mostly, the use of a shell and the degree of elaboration.

**Computer Combat Model with Map—** Computerized version of "Commercial Manual Wargame."

**1.** *Similar to Manual Model with Map.*

**2.** *Ideal Model for Senior Decision Maker.* Powerful model with easiest user interface.

**3.** *Can Be Easiest to Use (If Done Right).*

**4.** *Very Expensive to Create and Maintain.*
**5.** *Highly Interactive.*
**6.** *Forces.* Can handle larger number than manual model.
**7.** *Movement.* Can be more elaborate than manual model.
**8.** *Combat.* Can be more elaborate than manual model.
**9.** *Map Display.* Can be more elaborate than manual model.
**10.** *Rules of Use Built In.*
**11.** *Time Required to Create.* 2,500–12,000 hours. This can be done using a workstation or high-end, PC-based development system. If the coding is done in *C*, that code can be ported to other machine environments (Unix).

**Computer Combat Model Without Map—** The infamous, traditional, ''black box'' model. Or something very similar to it.

**1.** *Similar to Manual Model Without Map and Spreadsheet Combat Model.*
**2.** *Very Expensive to Create and Maintain.*
**3.** *Can Be Highly Interactive.*
**4.** *Forces Can Be as Large as You Want.*
**5.** *Highly Iterative, Unless It Chews Up Too Many Cycles Doing Its Basic Calculations.*
**6.** *Movement, but Usually Abstracted.*
**7.** *Combat, Usually Abstracted.*
**8.** *No Map Display.*
**9.** *Rules of Use Built In.*
**10.** *Time Required to Create.* 1,500–12,000 hours. In practice, some of these projects have gone on for years, apparently acquiring a life of their own.

**A QUICK CHECKLIST:** Important Things to Remember When Creating Models, Simulations, and Games:

**1.** Pay *very* close attention to the user. If he can't use the model, it may be the last one you'll do for him. (Not generally the case for Department of Defense contracts.)
**2.** Don't exceed your capabilities. Try to do more than you're capable of, and you'll make a mess of it.
**3.** A modest success looks better than a grand failure. Know your limits. Start with simple task. This is particularly essential if it is your first effort.
**4.** Use previous examples. There's no need to reinvent the wheel. Good ideas and techniques can be constantly recycled.

**5.** You can never test enough. Testing should use the same procedures applied to software. There are several levels of testing.

> **A.** *Unit Testing.* Test individual rule for soundness. Example, test the map for completeness and correctness.
> **B.** *System Testing.* Test rule along with other rules that it normally operates with. Example, test map with movement and combat rules.
> **C.** *Integration Testing.* Test all the rules together to see that all parts fit correctly.
> **D.** *Validation Testing.* Test the entire system to see that all user requirements are met.
> **E.** *Acceptance Testing.* User tests to see that all requirements are met.

**6.** Experience makes a big difference. Life experience (someone who is careful, persistent, and thorough) is valuable, more valuable than a lot of experience with using games. The techniques of modeling are simple. But they must be followed carefully in order to work. Carelessness can easily be fatal.

**7.** The model is never completed. You will always be coming up with new techniques. Users will continually come up with new requirements. Much the same experience as with software products in general. Bugs will continue to appear, although of a less severe nature over time. Users will almost always require enhancements and other modifications. Use of the model will point toward more efficient ways of doing things and will indicate that it is cost-effective to make revisions.

## USE OF HISTORICAL DATA IN MODELS, SIMULATION, AND GAMING ✷✷

It's often a less than perfect fit between historical data and modeling requirements for military professionals. There are numerous techniques that can ease the process. Here's a check list. If you run into a snag while trying to develop a game, run through this list again and again if need be.

### WHY USE HISTORY?

In a military environment, it's still often necessary to defend the use of the historical approach. Here are some good points to make to yourself or your boss.

**1.** History repeats and paraphrases. An examination of the historical record shows that progress is incremental, rarely revolutionary. You can see the future in the past.

**2.** History validates. Building historical models allows you to validate techniques used in models. This enables you to try new techniques freely, as you always have a quick means of checking the new technique's validity.

**3.** History provides models for models. Every historical situation is a potential model, complete with validation proofs, that you can use for comparing to other situations that have not happened yet. The 1991 Gulf War will be used for many years as the baseline model for other hypothetical situations.

**4.** More entertaining than calculus. The "game" element should not be underestimated. Anything that encourages use makes the model more useful.

## SHORT HISTORY OF MILITARY MODELING AND SIMULATION

In defending the use of historical wargames, it's often useful to have a quickly recapitulated history of wargames and their success in the past.

**1.** Chess as a battle model (antiquity to 18th century). Chess was originally an accurate model of pre-gunpowder combat.

**2.** Kriegspiel (19th century). Germans were the first to adapt chess model to more recent combat developments (1700s and 1800s).

**3.** Free Kriegspiel (late 19th century, early 20th century). Combined recent combat experience of armed forces and guidance of combat experienced officers to supplement extensive rules and play mechanics. Made for faster moving games. Major weakness was breakdown of system validity as officers and armed forces combat experience grew old and stale.

**4.** Kriegspiel versus the bean counters (1940–1965). The operations-research crowd threw out the baby with the bathwater when they applied their talents to combat modeling.

**5.** Kriegspiel meets rock and roll (1965–1980). Historical simulations came back with the baby-boom generation. These caught the attention and loyalty of many junior officers who bought, played, and eventually began designing them.

**6.** Kriegspiel goes electric (1980–present). Microcomputers enabled models to be created more easily and cheaply.

## SIMPLICITY AND EFFECTIVENESS OF HISTORICAL MODELS FOR INSTRUCTION AND RESEARCH

Here are the key points to be used for research or instruction if you are dealing with wargames.

**1.** *Visibility.* You can see all aspects of what's happening.

**2.** *Validation.* You can see if, and how, the model accurately simulates the historical event. Enables you to freely test new ideas without having to worry or argue about validity. Historical proof is always handy.

3. *Versatility*. Useful for training and testing. "What if?" possibilities enable research and analysis to be performed.

### DATA SOURCES FOR CREATING WARGAMES

Afraid of not being able to do the research needed to create a game? Here are the key points to constantly keep in mind.

1. Same as for a history book. Start with a good secondary source bibliography, then work down to primary sources as needed.
2. More analytical data needed. Books with lots of charts and tables are more useful.
3. Electronic data bases for contemporary subjects. These are becoming more prevalent and useful than paper sources.
4. Create a data model first. As in my books *How to Make War* and *A Quick and Dirty Guide to War*.
5. Model building will smoke out more data needs. A model is working when it begins to raise questions. This often happens before the model begins providing any answers.

## POLITICS IN MODELING ✱✱

A military model must deal with many items that are not, strictly speaking, military. For the most part, these nonmilitary items have to do with various "political" issues. We are talking about the existence of political considerations within wargames, as well as political influence on the way wargames are designed. Examples include:

**Patronage and Battlefield Performance.** Patronage is always with us. Many countries, including the United States, must cope with political interference in purely military decision making, on questions such as what weapons to purchase, how many, and in some cases, where to station them. A more critical problem arises when senior military commands are handed out on the basis of political, not military, competence. Battlefield performance is a serious issue only during wartime. Sometimes even then, it is vague. The combat-experienced people are often unable to prevail over those who simply create experience in their imaginations and memos.

**The Mitrailleuse Factor.** The Mitrailleuse was the French secret weapon of 1870 that was so secret no one in the French Army knew how to use it. There are a lot of weapons like this still around. Sometimes no one knows how to use the weapon simply because it's unique, and there's no past experience to use as a guide. The "politics of secrecy" in most military establishments frequently keeps valuable information from their own troops. This was noted as recently as the 1991 Gulf War.

**The Red Badge of Courage Factor.** Combat experience, or at least military service, imparts an invaluable insight into obscure aspects of military operations. Noted military authors don't catch the essence of the chaos on the battlefield as well as novels like *The Red Badge of Courage* (written by a man who was never in combat, and who was hailed as one of the few noncombatants who managed to capture the feel of the battlefield). These differences in perceptions are often the explanation for otherwise inexplicable historical events.

**Learn to Live with Paper Bullets.** Physical death is a wartime danger. In peacetime, the normal state of soldiers and their suppliers, paper bullets can cause career-threatening injuries. Soldiers have to live with this danger for most of their careers. Even in combat, people not getting shot at by real bullets are always subject to the paper variety. Paper bullets must be modeled if you want to have accurate models of military behavior.

**Unsinkable Aircraft Carriers.** If your model is really good, it may destroy some zealously held beliefs. This may cause you some problems, as the model may be seen as more expendable than the beliefs.

**Unquestioned Optimism.** The guidance you receive from your clients may cause you to create a model with unrealistically optimistic goals or parameters. A good model builder accepts the burden of disabusing the client of his myths. The goal of a model is to reveal the truth, not perpetuate false perceptions.

**Your Army Won't Go to War with Itself.** Armies, and frequently model builders, make the mistake of modeling their opponents as too similar to themselves. Your potential opponent must be modeled accurately. Think like the enemy, which is not as easy as it may appear. Use teenagers or others with few preconceived ideas to model opponent forces.

## END NOTE ✱✱

This is the end of the book. I could go on and on, but my editor only gave me so many pages. I talked her into giving me some more, but for a subject so vast, it's never enough. So how you do you move on beyond what I have presented in these pages? The answer is simple and obvious. You go and play the games, or study the games, and perhaps you'll get really ambitious and create your own games and simulations.

Wargames are a new form of media that have been lurking in the shadows for thousands of years. It's new because in the past there were never a large enough number of highly educated people to form a true community of gamers. This has changed in thc last thirty years. Wargames are very much a side effect of widespread higher education. Computers have made the games even more accessible.

Best of all, gamers have learned how to see beyond games on military conflict. My own interest was from the social history angle, and that explains why I often (with varying degrees of success) emphasize the nonmilitary aspects of the wargames I have dcsigned. Yet games of war are but the tip of the iceberg in gaming. There's politics, economics, and every aspect of human life you can think of, and these belong in wargames also. That games like *SimCity* and *Civilization* appeal to wargamers is an indication of this trend.

Carry on.

# APPENDICES

Someone is interested in wargames because someone is interested in history. The playing of wargames and the study of written material go hand in hand. Quite normally, the interest in one begets an increasing interest in the other. It is common for someone to have read something about a particular historical event and then to seek out a game on a similar event: You then use the game to enlarge your insight and understanding of the historical situation. Conversely, the gamer is just as likely to pick up a game without really knowing anything about the historical situation and, using the experience with the game, read more and study more on the subject.

There are primarily two kinds of research materials the hobbyist can use: general surveys and specific sources. General surveys are quite important, as they give a shape to one's research. The specific sources are needed when one gets down to a particular game subject.

Interest in most of the specific sources, as well as getting interested in a particular area, are usually the result of reading a recent book on the subject. It is new books such as these that get into most people's hands, although it often happens that someone stumbles across an old classic and becomes inflamed about, say, the Thirty Years' War. Generally, though, it is more likely that a spate of recently published books on a given war will generate far more interest than the chance encounters with older works.

History is constantly being revised. Otherwise all historians would pretty much be out of business. Your best bet on keeping with what is currently being produced is to frequent your local bookstore. If there is no bookstore in your town, contact the History Book Club or the Military Book Club through the mail. If you are not satisfied with what is being displayed in the bookstore, ask to look through the store's copy of *Books in Print* and/or *Paperback Books in Print*. The store should be able to get any book that is listed in these two volumes for you.

There are a few basic sources that will aid you in maintaining a sense of where you are in military history. Probably the single most valuable work is Dupuy and Dupuy's *Encyclopedia of World Military History*. This is a massive survey of warfare from prehistory to the present and gives considerable detail on campaigns, battles, and military systems in general. For starters, before you get down to specific cases in any particular period, you should concentrate on the survey-type works. David Chandler has done three excellent survey books. The best known is probably *The Campaigns of Napoleon,* but he also did two on the early 18th-century general, Marlborough: *The Art of War in the Age of Marlborough* and *Marlborough as Military Commander*. An excellent series of atlases is to be found in Esposito and Elting's *A Military History and Atlas of the Napoleonic Wars* as well as Esposito's two-volume *The West Point Atlas of American Wars*.

It is quite common for the gamer to specialize in one or a few particular periods. This is advisable. It increases one's enjoyment of both the books and the games on that era. One of the more popular periods of wargaming is the contemporary period. This is history in the making, and there are a large number of periodicals addressing this subject in considerable detail. A good research library will have copies of the following to peruse, from which you can obtain current subscription information:

*The International Defense Review* is a gold mine of excellent, up-to-date information on contemporary military affairs that is not likely to be found anywhere else.

*Infantry* magazine (Box 2005, Fort Benning, Georgia 31905), *Armor* magazine (Box 0, Fort Knox, Kentucky 40021), and *The Field Artillery Journal* (Fort Sill, Oklahoma) are all U.S. Army publications that provide very current information and contemporary and historical items within each magazine's area of interest. *Military Review* (USACGSC, Fort Leavenworth, Kansas 66027) is the army's highest-level professional journal and contains many "think" pieces as well as some good items on tactics, strategy, foreign armies, and military history.

*Army* magazine (1529 Eighteenth Street N.W., Washington, D.C. 20036) features contemporary and historical articles of interest to the Army Reserve officer (for whom it is published). *Air Force* magazine (1750 Pennsylvania Avenue N.W., Washington, D.C. 20006) is similar to *Army,* although it usually has more meaty articles. The *U.S. Naval Institute Proceedings* (Annapolis, Maryland 21402) is an excellent periodical for naval affairs.

A host of other foreign (many in English) periodicals (and books) can be obtained from Sky Books International, 521 Fifth Avenue, New York City 10017).

## * * MY GAME BIBLIOGRAPHY * *

I have authored over 100 published simulations of historical and current events. I'm constantly asked, "Did you design this one, or that one?" and sometimes even I'm not sure. So I put together a list. Most were published by SPI, although at least five other publishers are also represented. A dozen or so are still in print; the rest are collectibles. Consider this an example of what one person can do if he applies himself.

### TACTICAL LAND-COMBAT SIMULATIONS

**Men-At-Arms** (1990)—Covered combat from the dawn of history to the 16th century.

**Tactical Combat Model** (1985)—Vehicle/crew, served weapon-level game of antivehicle operations. Emphasis on future robotic munitions and other standoff weapons. Prepared for a corporate client for in-house use only.

**Agincourt** (1978)—A very accurate rendering of thc 14th-century battle. Shows how traditional histories can be upstaged by well-done models. We found that the battle could not have been fought as it is usually described, although Keegan's *The Face of Battle* came close to our conclusions.

**Firefight** (1976)—Done under contract for the Infantry School. To be used for training platoon leaders and senior NCOs. Tactical-level; units represented vehicles and fire teams. Contract allowed for commercial version, which became a best-seller.

**Strike Force** (1976)—Simple, tactical introductory game, designed to accompany *Firefight* to introduce wargaming to new users.

**Panzer '44** (1975)—Tactical armored warfare in Western Europe, 1944–45. Uses same game system as *Mech War '77*, demonstrating the connection between historical and contemporary modeling techniques.

**Mech War '77** (1975)—An update of *Red Star/White Star*, using a new game system. More attention paid to command control and the nuances of the more recent weapons.

**Tank** (1974)—Vehicle-to-vehicle combat in 20th century. Wide variety of equipment, organization, and doctrine covered. Leadership and command control covered.

**Patrol** (1974)—A rural version of *Sniper*, with many new wrinkles.

**Scrimmage** (1973)—A very tactical game of football. Was very helpful in solving some of the game-design problems encountered in tactical games that dealt with firearms.

**Sniper** (1973)—Man-to-man operations in urban environment. Veterans of Tet were able to accurately replay their actions in Hue, etc. Paid attention to morale and leadership problems. Covered wide variety of 20th-century historical and contemporary situations.

**Desert War** (1973)—Tactical armored warfare in western desert (North Africa) during 1941–43.

**Kampfpanzer** (1973)—Tactical armored warfare in Western Europe, 1940–41.

**Combat Command** (1972)—Tactical armored warfare in Western Europe, 1944–45.

**Outdoor Survival** (1972)—Simulation of an individual surviving in a wilderness. Went on to sell for many years because RPG players used it as part of their home-brewed scenarios. If I'd only known . . .

**Red Star/White Star** (1972)—Designed at the request (and support) of staff of the Infantry School. A tactical-level simulation (maneuver units represented platoons) of combat in the central front. Stressed use of ATGM, coordination of mech infantry, armor, and artillery. This was the first commercial simulation on modern combat. Very successful commercially.

**Grenadier** (1971)—Tactical-level warfare (platoon-size units as playing pieces) during the 17th, 18th, and early 19th centuries.

**PanzerBlitz** (1970)—Tactical-level combat on the Russian front (1941–45). Over 300,000 sold.

**Deployment** (1969)—Tactical-level warfare (platoon-size units as playing pieces) during the 17th, 18th, and early 19th centuries. I began designing this one in 1964 while I was in the army (in Korea, of all places). Probably the first wargame ever designed in Korea. Undoubtedly the first wargame of this period ever designed in Korea. A dubious record at best.

## OPERATIONAL-LEVEL SIMULATIONS

**Light Infantry Division** (1985)—A game simulation of light-infantry operations in the Persian Gulf area. Part of an Army War College project on light infantry. Commercial version of this simulation was published later, after the design was considerably rehandled.

**The Drive on Metz** (1980)—The demonstration game found in this book.

**Fifth Corps** (1980)—The first of a series of games on a more detailed scale than *The Next War*. Followed the same general approach as *The Next War,* but the rules were simpler and more refined. Each game was independent, but could be linked with adjacent games. The series would, when finished, produce a game that, in map size, would be more than three times the size of *The Next War*. The latest title (*Donau Front*) in this series was published in 1989, just before the Berlin Wall came down.

**NATO Division Commander** (1980)—Concentrates on problems a division commander would have in contemporary combat in Europe. Intelligence, fatigue, command control, logistics, leadership (immediate subordinates' personalities and capabilities), and a host of other factors covered. Has version where player faces a controller who plays opposing forces and manages intelligence operations.

**Berlin '85** (1980)—A battalion-level (most units represent a battalion or regiment) game of Warsaw Pact attack on NATO Berlin garrison. Highly detailed display of militarily significant aspects of West Berlin on map. Combat mechanics relatively simple. A very active tactical game. Was denounced in *Kraznaya Zvesda* in 1986 for implying that the Russian Army would ever attack NATO garrisons in West Berlin.

**Bulge** (1979)—A fast-playing game of the Battle of the Bulge (1944).

**Brusilov** (1978)—The last major Russian offensive in 1916.

**The Next War** (1978)—A large-scale operational-level game covering the Baltic-to-North-Italy area. Included naval and air operations. A wide variety of doctrinal, operational, political, and technical factors were modeled in the game. Fatigue, command control, supply, tempo of operations, and many others.

**Fulda Gap** (1977)—Operational-level game of potential contemporary operations in the Fulda Gap area. Covered a wide range of factors a commander would have to deal with.

**Panzergroup Guderian** (1976)—The German drive on Smolensk in 1941. Emphasis on leadership, unknown troop quality, and blitzkrieg tactics.

**Wurzburg** (1975)—An operational-level game (scale 1:100,000) of contemporary operations around the German city of the same name. Denounced by Burgermeister of Wurzburg for implying that his fair city could be fought over and possibly nuked. Game uses battalion-sized playing pieces. Was used by at least one 7th Army brigade commander for planning operations. Game was originally designed over a weekend to generate scenarios for *Firefight*.

**Oil War** (1975)—Covered hypothetical invasion of Persian Gulf area in response to a number of different threats. Showed difficulty of such operations.

**The East Is Red** (1974)—Division-level rendering of possible Sino-Soviet operations in Manchuria. Demonstrated both sides' unique advantages and disadvantages.

**Combined Arms** (1974)—Battalion-level combat in the 20th century with emphasis on interaction between the principal combat arms.

**Operation Olympic** (1974)—Proposed Allied invasion of Japan in late 1945. Solitaire (only Allied player active).

**Napoleon at Waterloo** (1973)—Fast-playing game of 1815 battle.

**Destruction of Army Group Center** (1973)—Russian offensive during summer of 1944.

**Panzer Armee Africa** (1973)—The war in North Africa (1941–43) with emphasis on movement and logistics.

**El Alamein** (1973)—Set-piece North African battle in 1942.

**Ardennes Offensive** (1973)—Battle of the Bulge (1944), with emphasis on operational maneuvers.

**Battles of Bull Run** (1973)—1861 battle in American Civil War with emphasis on command control and communications.

**Sinai** (1973)—Series of scenarios (operational level) on Arab-Israeli wars. Design finished when October war broke out. Accurately predicted course of war in "hypothetical '70's War." Showed value of predictive games based on historical trends. In published version included '56, '67, '73, and "future" wars. Received research assistance from members of Israeli UN delegation, who were anxious to find out what was going on and how the war would develop.

**Year of the Rat** (1972)—NVA 1972 offensive. Original design by John Prados; I had to completely redesign it to get it to work. Game featured a set of G2 rules that realistically re-created uncertainty of NVA strategy. Game demonstrated value of U.S. firepower (B-52s, etc.) and mobility. Got high praise from players with access to classified details of actual operations.

**France '40** (1972)—Operational-level simulation of German operations in France during the spring of 1940.

**Wilderness Campaign** (1972)—Civil War battle, 1864. Emphasized difficult terrain and command-control problems.

**Moscow Campaign** (1972)—German drive from border to Moscow, 1941.

**Breakout and Pursuit** (1972)—Breakout from the Normandy beachhead and drive toward Paris, 1944.

**Turning Point** (1972)—Stalingrad Campaign, 1942–43.

**Kursk** (1971)—Battle on Russian front in 1943. Extensive treatment of "what ifs" in timing of German attack and state of Russian preparations.

**Lost Battles** (1971)—Regimental-level operations in Russia, 1943–44, with emphasis on differences between German and Russian methods of fighting.

**Bastogne** (1970)—Battle of the Bulge (1944), with emphasis on road-net problems.

**Anzio Beachhead** (1969)—Multi-level game of the Italian Campaign in 1944.

**Crete** (1969)—German airborne operations in Crete, 1941.

**Italy** (1969)—Multi-level (operational/tactical) game of fighting in Italy (1943–45).

**Korea** (1969)—Operational level, 1950–51.

**1918** (1969)—Final German offensive in the west.

**Tannenberg** (1969)—Opening battles in the east, 1914.

**Normandy** (1969)—Regimental level treatment of Allied 1944 invasion of Europe.

**Leipzig** (1969)—Operations in Central Europe during Napoleonic Wars (1814).

**1914** (1968)—Opening operations on the western front in 1914.

## STRATEGIC-LEVEL COMBAT SIMULATIONS

**Hundred Years War** (1992)—A multi-player computer-run game of the titanic struggle between England and France in the 14th and 15th centuries. Run on the GEnie network and capable of supporting over 300 players per game.

**Empires of the Middle Ages** (1980)—Operations in Europe from A.D. 790 to 1452. Won awards for its elegant treatment of political, economic, and military issues on a strategic scale.

**Revolt in the East** (1976)—Theater-level game demonstrating problems Russians would have dealing with multiple uprisings in East Europe.

**War in Europe** (1976)—Combination of *War in the East* and *War in the West* with additional rules and mechanics to link the two theaters.

**Russian Civil War** (1976)—Military/political operations in Russia, 1917–21. One of the few games where assassination played a significant role.

**World War I** (1975)—A small, compact treatment of the 1914–18 war.

**Battle for Germany** (1975)—Final battles in Europe. Three sides, Western Allies, Germans, and Russians.

**Global War** (1975)—All of World War II, worldwide.

**World War Three** (1975)—A global game demonstrating the importance of sea power and strategic position. Covered amphibious and economic warfare.

**Invasion America** (1975)—An exercise in continental defense. A bit on the science-fiction side, but using contemporary weapons and force models. Postulates the United States defending North America against a variety of invaders.

**NATO** (1973)—Division-level simulation of Europe from Baltic to Alps. Used as manual base game for *MTM* at USAWC.

**War in the West** (1976)—Western Allies–German war (1939–45) in great detail on huge maps.

**War in the East** (1974)—Russian-German war (1941–45) in great detail on huge maps. Head of Russian UN delegation came by to personally pick up a copy of this one. He said they had copies of all our other games, but were particularly interested in this one.

**American Civil War** (1974)—Emphasis on leadership, movement, and logistics.

**Solomons Campaign** (1973)—Air, land, and naval operations in the South Pacific in 1942–43.

**World War Two** (1973)—Russia and Western Europe and North Africa, 1939–45.

**Franco-Prussian War** (1972)—Operations between France and Germany in 1870 with emphasis on limited intelligence.

**American Revolution** (1972)—Entire war (1776–81), with emphasis on leadership.

**USN** (1971)—Relatively simple game of theater-level operations in the Pacific, 1941–43.

**Strategy I** (1971)—Using hypothetical map, contains variety of rules and scenarios allowing creation of many typical strategic situations.

**Barbarossa** (1969)—Russian-German war (1941–45) with emphasis on movement and maneuver.

## NAVAL COMBAT SIMULATIONS

**Victory at Sea** (1993)—The naval war in the Pacific during World War II. A computer game for PCs.

**The Fast Carriers** (1975)—Carrier operations from World War II to the present. Showed the development of doctrine in response to changes in weapons and equipment.

**Sixth Fleet** (1975)—Naval game showing ship and aircraft operations in the Mediterranean during the 1970s. Demonstrated importance of "combined arms" (surface and submarine ships, land- and carrier-based aircraft).

**War in the Pacific** (1975)—Huge game of World War II in the Pacific. I did the production system, which was a game in itself.

**Wolfpack** (1974)—Submarine operations in the Atlantic, 1942–45. A solitaire game, with the Allies as the only active player.

**Frigate** (1974)—Tactical, ship-to-ship combat during the age of sail and cannon.

**CA** (1973)—Surface warship combat in the Pacific, 1942–45.

**Jutland** (1967)—Naval battle in North Sea, 1916.

## AIR COMBAT SIMULATIONS

**Foxbat and Phantom** (1973)—Modern air combat, plane to plane. Demonstrated inadequacies of MiG-25 as an air-combat aircraft before this fact became known, or fashionable to admit. Demonstrated the basic rules of air combat, most notably the importance of surprise, superior position, and the vastly differing capabilities of aircraft sensor and weapons systems.

**Spitfire** (1973)—Fighter combat in Europe, 1939–45.

**Flying Circus** (1972)—Air warfare, tactical, 1914–19

**Flying Fortress** (1969)—Strategic bombing campaign over Europe, 1943–45.

## POLITICAL CONFLICT

**Canadian Civil War** (1978)—Political infighting among Canadian provinces and federal government. Sold over 10,000 in Canada alone.

**Plot to Assassinate Hitler** (1976)—Maneuvering of the anti-Hitler conspirators and the Gestapo during World War II.

**Origins of World War I** (1972)—Similar to *Origins of World War II,* but for World War I.

**Origins of World War II** (1971)—Political simulation of the pre–World War II maneuvering in Europe.

**Chicago-Chicago** (1970)—Political-action, media-manipulation game based on the 1968 riots at the Democratic Convention.

**Up Against the Wall, Motherfucker** (1969)—The game of the student takeover of the Columbia University campus in the spring of 1968. I was a student there at the time, and although I didn't participate in the action, several friends did. Some of these lads worked on the school newspaper (*The Columbia Spectator*), and they asked me to do a game on the event for the first anniversary special issue of *The Spectator*. I agreed to do the game as long as they published it under the name of my choosing. Although I was an elderly twenty-five years old at the time (attending Columbia under the G.I. Bill), I still had the spirit of the '60s. The game title caused a bit of a stir, but the game was a simple and accurate representation of the power struggle that went on in the spring of '68.

## FANTASY AND OTHER FICTION

**Demons** (1980)—A game involving magic, set in ancient Armenia. Did this one on a dare.

**Wreck of the Pandora** (1980)—Science-fiction game about surviving a wrecked spaceship full of nasty biological specimens.

**Dallas** (1980)—Role-playing game based on TV show of the same name.

**TimeTripper** (1980)—Science-fiction game of Vietnam-era soldiers wandering through combat in past (and future) battles.

You can't use wargames without the games themselves, and the experience can be further enhanced with a host of supporting items and services. Here you find out how to get this stuff, or at least learn that a lot of it actually exists. Because this book is likely to stay in print for many years, I will refrain from giving the names and addresses for specific vendors here, as many companies in this business come and go. Where I do give a name and address, it is either because there is little choice (as in direct-mail vendors for people with poor access to retail outlets) or because the company in question has been around for a while and is likely to still be in business 10 years from now.

**Retail Outlets—** This is the easiest way to get your hands on wargames. Most major metropolitan cities have stores dedicated to games and some of these are stocked with a wide selection of wargames (both computer and manual). Most also carry wargaming magazines and related items. Check your Yellow Pages.

**Direct Mail Outlets—** If you can't get to a store, try direct mail. Most carry both manual and computer wargames. The direct-mail outfits also discount the games, so you save a few bucks. Alas, you can't look at the box or an in-store demo, but service is usually quite fast, with overnight delivery available from most companies. You can get the numbers of current direct-mail vendors from gaming magazines. But if you're really stuck (and can't get your hands on any magazines), try Cape Cod Connection (800-328-9273) or Chips & Bits (800-753-4263). For Canadians (and those in the United States also) there is Games by Mail (613-523-3699). Games by Mail also sells used and out-of-print games, as well as out-of-print errata for them.

**Introductory Games—** GDW has, for over a decade, sold a low-cost and very simple introductory wargame to get people into the hobby. The introductory game changes from time to time and is usually carried in stores.

## WARGAMES AS COLLECTIBLES

Wargames, in many cases, increase with value as the years go by. The combination of games going out of print and high-income gamers has created a collectibles market. This all began in the late 1970s when the game conventions began featuring auctions. At first, this was an opportunity for gamers to sell off games they had already played, or were simply unhappy with. But from the very beginning there were out-of-print games offered, and even then these games tended to sell for more than their original list price. As gamers got older, and wealthier, and more games went out of print, the value of some of these games escalated considerably. For example, as of early 1992, these are some of the prices you would encounter. In addition to the publisher and year of publication, the three prices shown are the original price, the original price in 1992 adjusted for inflation, and the current price range of the game.

*Wacht am Rhein* (SPI, 1977, $20/$46/$300–$500)

*War in the Pacific* (SPI, 1978, $30/$65/$300–$600)
*Campaign for North Africa* (SPI, 1979, $30/$58/$150–$200)

These first three games represent the most sought after class of games, the "monster games." These are characterized by having two or more full-size maps and over 1,000 playing pieces. They sell for four to ten times their original (inflation-adjusted) price.

*Empires of the Middle Ages* (SPI, 1980, $18/$31/$120–$200)
*Kharkov* (SPI, 1978, $10/$21/$30–$40)

These next two games are examples of standard-size games. *Empires of the Middle Ages,* an award-winning and much sought after (and played) game gets the highest prices in this category. The current ratio of four to six times its inflation-adjusted original price will no doubt go up. *Kharkov* is at the bottom of the range, being a game originally published in *S&T* and then later released in boxed format for retail sale. Even so, it still sells for up to twice its original (inflation-adjusted) price.

*Crimean War Quadrigame* (SPI, 1978, $14/$30/$80–$100)

SPI published 16 Quadrigame sets (containing 64 separate games) between 1975 and 1979. Many of the subjects covered were obscure, and these games represent the only time these events ever made it into wargame format. All of these games are out of print, and demand is brisk. While the Crimean War quad is one of the most sought after, and sells at two to three times its original (inflation-adjusted) price, most of the other quads sell for at least their inflation-adjusted price and many for twice that.

The current prices given above are the "mint"-condition copies of these games. That is, games as they originally appeared in the stores, with their original plastic shrink wrap intact. In other words, unplayed and untouched by human hands since manufacture. Beware, however, that some unscrupulous collectors are taking a played copy of the game and, at worst, filling the box with junk and then shrink-wrapping it again. Pretty neat scam, as they are unlikely to be found out unless the buyer decides to play the game. Note that an already played game (with the die-cut playing pieces punched out) reduces the value of the game 20–30 percent. If the game components are in poor condition or some are missing, value can be reduced a further 10–20 percent. Thus, a played copy of *War in the Pacific* would go for $200–$500. If some components are missing and the game is beaten up, the value goes down to $160–$450.

There aren't that many collectors, only 1,000 or so. But their number increases each year. Moreover, some of these games can still be found in out-of-the-way stores or flea markets, usually at very low prices. But that's part of the appeal, and attraction, of collectibles. However, many gamers seek out these games to actually play them. In order to have it both ways, some collectors will make color photocopies of the map and playing pieces (and then mount and cut out the new copy of the playing pieces) and thus retain the value of their collectible and still be able to play it. Some collectors are selling these copies, although this violates copyrights and is illegal.

The games I have mentioned are all SPI games because, quite frankly, those were the only ones I had original price and publication-date data for. However, SPI games are the largest single class of collectible games you will encounter, if only because SPI published so many. TSR currently owns the copyrights for most SPI games and republishes some of them each year. This will reduce the value of those titles somewhat, but not entirely. There are games of other publishers, particularly small publishers of the late 1960s, that have value similar to *War in the Pacific*. SPI games do, as a group, have the highest collectible value. As with all collectibles, their value can only go up, since only 5,000–10,000 copies of each game were published. Unlike stamps, comic books, baseball cards, and the like, where collectors will buy large quantities and hold them until the value increases, nearly all SPI games were bought to be used. No one was thinking about a collectibles market. The mint-condition games are usually those bought by a gamer who ran out of time to play, or even look at, the game.

Collectibles are where you find them. Some vendors sell them, at prices somewhat above the prices discussed above. The game conventions have auctions and swap sessions. The computer services (below) enable gamers to buy and sell electronically.

Speaking of computers, there is also a market in computer-wargames collectibles. This market is a bit different. You need a specific type of computer to play a computer wargame on. Ten years after a computer wargame is published, there are only few gamers who still have the type of machine needed to play the old game. There is also less in the way of physical components in a computer wargame, usually just one or more diskettes, a box, and a small manual. As a result, the collectibles market in computer wargames is not developing into quite the robust creature the paper-wargames market is. Currently, 10-year-old computer wargames are lucky to bring half their original list price (without any adjustment for inflation). There's some value in that ten-year-old computer wargame that cost you $20 in 1980, but not nearly as much compared to a 1980 copy of *Empires of the Middle Ages* that originally cost $18.

## MAGAZINES

Largely because wargamers are such a well-educated and literate lot, there are a disproportionate number of magazines serving the hobby. Most are edited and published by wargamers. Quality is generally high. All of these magazines can be obtained by subscription from the indicated publisher. Many of them are available in bookstores, hobby stores, and toy stores that carry wargames.

**Command Magazine** (published by XTR Corp.); bimonthly. Similar to *Strategy and Tactics*. A history magazine with a game in it.

**Courier.** For miniatures wargamers.

**Computer Gaming World** (published by Golden Empire, 130 Chaparral Ct. 260,

Anaheim Hills, CA 92808); monthly. The premier magazine of reviews and analysis of computer games. Good coverage of computer wargames, by wargamers.

**Fire and Movement** (published by Decision Games); bimonthly. *Fire and Movement* is primarily a magazine of reviews. It has done very well at this since there are so many new games coming out. The reviews range from very brief mentions (''capsule reviews'') to lengthy multi-article pieces on one game. It's well worth reading if you want to get a better idea of what some of the games coming out are like.

**The General** (published by the Avalon Hill Game Company, 4517 Hartford Road, Baltimore, Maryland 21214); bimonthly. It alone among all of the other wargame publications is devoted solely to Avalon Hill games. Most of the articles have to do with tactics, variants, additional scenarios, and the play of Avalon Hill games.

**Moves Magazine** (published by Decision Games); bimonthly. *Moves* is the magazine of gaming theory and technique. It covers all games in the hobby and has articles of opinion, theory, technique, reviews.

**Simulations Online Magazine.** Prepared by a bunch of wargamers on the GEnie online service (the ''Games RT'') and distributed in electronic (ASCII) format over many other systems. If you are a modem junkie, you should be able to find it. Lots of good stuff for the wargamer.

**Run 5.** House organ of the Strategic Simulations Group (SSG). Carries material on SSG's own computer wargames.

**Strategy and Tactics** (published by Decision Games); bimonthly. It contains at least two military-history articles and a complete game (usually with a 22-by-34-inch map sheet, at least 200 playing pieces, and eight to twelve or more pages of game rules). One of the history articles is on the same subject as the game, and in addition there are reviews of games and books, other features, and a fairly lengthy column detailing what is going on in publishers' game R&D departments. Also, a reader-response (feedback) section.

**Strategy Plus.** British version of *Computer Gaming World.* Features shorter reviews, done in a uniquely British style.

## GAME ASSISTANCE PROGRAMS (GAPs)

These are computer programs to assist in the play of manual games. GAPs keep track of the detail of complex manual games. There was never a significant market for these items, although a few were published in the 1980s. Several can be found on electronic bulletin-board systems (BBSs) and the national on-line services.

## COMPUTER SERVICES

CompuServe and GEnie are two publicly available computer services that you can reach with a PC and a modem (over your telephone line). Both have very active wargaming sections where gamers can exchange information and download (transfer) to their systems all manner of material. This includes text files explaining various aspects of games (how to play, reviews, and so on) or a large number of shareware (you pay for it if you like it) computer wargames. The shareware games are, in general, not as good as the commercial games, but many shareware products are quite worthwhile. You can also download demonstration versions of commercial wargames. This is a good way to get a look at a game before you play it. Call 800-638-9636 if you want to sign up for GEnie. CompuServe has start-up kits in most computer-software stores.

Many independent bulletin-board systems (BBSs) also have facilities to run games on-line. Get a copy of *Computer Shopper* (available in most large magazine stands) for a listing of some BBSs in your area. All you need is a PC and a modem. Not as slick as the commercial services, but there are thousands of them all over the country (and the world, even Russia). Watch out for calling any of them long distance. You tend to lose track of time, but the phone company's billing system doesn't.

## ORGANIZED GROUPS

Most large metropolitan areas have one or more groups of organized (or semi-organized) gamers. Finding the group can be tough, though. Some of the game magazines list groups regularly or from time to time. Computers bulletin boards (BBSs) often provide an opportunity to find such groups. Game conventions are another way to make contact.

## GAME DESIGN MATERIALS

For those who want to get into designing games, you'll need some raw materials (blank hex sheets, blank counter sheets, etc.). One of the best sources in Zocchi Enterprises (601-863-0215).

## ✱ ✱ PAPER WARGAME PUBLISHERS ✱ ✱

The publishers of wargames are a diverse lot. They all generally have one thing in common: The people producing the games are gamers themselves. Moreover, the management of many of the publishing companies also consists of hobbyists. This factor generates an unending stream of interesting and usually good (and always sincere) games.

As with most "industries," the wargame field is dominated by a handful of companies. At this point, Avalon Hill is by far the largest factor in manual-games publishing. Avalon Hill was the original publisher of wargames, and it never published that many, preferring instead to promote energetically the sales of the small number of games it had published. This was largely because it had the widest distribution system and could have developed a serious inventory-control problem if it had had too many titles in print. Avalon Hill also republishes new editions of its old games as well as other publishers' games that it has bought the rights to. Victory Games, basically an imprint of Avalon Hill, publishes games in the old SPI style, for which there is still a market.

Avalon Hill, because of its origins in the earliest days of wargames publishing, is practically the only company that manufactures games with stiff, mounted map boards. It is able to maintain competitive prices, even with this, because of the enormous volume it does with each of its games. Most other games are sold in volumes equivalent to those of hardcover books (5,000 to 10,000 units per title). Avalon Hill, on the other hand, can still sell more than 10 times that many units per title. This gives it economies of scale that are passed on to the consumer. In addition to its wargames, Avalon Hill puts out an expensive line of sports and nonhistorical "adult" games. Its in-house R&D (about half a dozen people) spends most of its time working on game-related projects. In the last 20 years, Avalon Hill has published nearly 100 games.

Below Avalon Hill, there are about a dozen other publishers that often publish fantasy and science-fiction games in order to make it possible to publish the less profitable wargames. Together, these smaller companies probably account for nearly half of all wargames sold. The smaller wargames publishers are:

*Australian Design Group*. A small operation down under that puts out a small number of high quality games.

*Clash of Arms Games*. Small outfit doing highly detailed wargames.

*Decision Games*. Current publisher of *Strategy and Tactics* magazine (each issue with a game in it). Also publishes games for distribution in stores.

*Fresno Gaming Association*. A new company dedicated to bringing out large, very detailed wargames.

*The Gamers, Inc*. A small outfit that publishes only wargames (a rarity these days).

*Game Designers Workshop (GDW)*. After Avalon Hill, the largest wargame publisher. GDW now publishes mainly fantasy and science-fiction titles to pay the bills.

*GMT Games*. A new publisher (1990) that has so far put out a number of high-quality and popular wargames.

*Games Research/Design* (*GRD*). This is a unique outfit. Its primary purpose is to keep in print the *Europa* series of interlocking games on World War II. The *Europa* series was begun by GDW in the 1970s, and the rights were bought by GRD in the late 1980s. The *Europa* series has a devoted following of enthusiastic military historians.

*Milton Bradley*. Yes, this is the same major game publisher. Outfits this size have, on occasion, attempted to enter the wargames field. Most soon discovered that it was much too small to be worth the effort. However, Milton Bradley has published several simple wargames that still appeal to the mainline wargamer. The games have stayed in print, and are available just about everywhere.

*Nova Game Design*. Noted mainly for its unique booklet-based air-combat games.

*Omega Games*. So far, has done games primarily on contemporary topics.

*Palladium Books*. Primarily a role-playing game publisher, but it has done several titles on modern historical military topics.

*Simulation Design, Inc*. Small company devoted mainly to American Civil War games.

*TSR* (*Tactical Studies Rules*). This originally started out as an operation to publish rules for playing wargames with miniatures but soon developed a game called *Dungeons and Dragons,* which was the first of the fantasy role-playing games. This in turn developed into one of the first truly ''mass appeal'' wargames. By virtue of this one product catching on, TSR was catapulted into the upper ranks of game publishers. Most of its other games have been less successful, traditional-type games. The most successful products TSR has published to date have been those that supplement and support *Dungeons and Dragons.* When TSR absorbed SPI in 1982, it continued to publish some of SPI's older titles, as well as some new ones.

*West End Games*. Originally a publisher only of wargames. Over the years, its output of wargames has declined (but not disappeared). Fantasy and science-fiction titles production has increased considerably.

*World Wide Wargames* (*3W*). A British transplant to California. So far, it publishes it only wargames.

*XTR*. The publisher of *Command* magazine (with a game in each issue), it also publishes a line of inexpensive games that are available in stores. XTR is one of the major proponents of ''alternative history'' games. Things like, ''What if the Germans or Japanese had won World War II, or the Confederates had won the Civil War,'' and so on. They published a game on the 1991 Gulf War that gave the Iraqis an excellent chance of winning. There is a market for this sort of thing, although it's an acquired taste.

There are several other publishers, most with only one title to their credit and likely to disappear when that one title disappears. There are also several other former wargame publishers that currently do only fantasy and science fiction (Mayfair, Steve Jackson Games), but, as they are run by wargamers, may again publish some wargames.

## ✱ ✱ COMPUTER WARGAME PUBLISHERS ✱ ✱

This is an even more diverse lot than the paper-wargame publishers. This is not a complete list, as there are even more outfits zipping in and out of wargame publishing on the computer end of things.

**Accolade.** Primarily does arcade-type games, but on occasion does a passable wargame title.

**Applied Computer Concepts.** A small outfit, not much output.

**Avalon Hill.** Long known for its paper wargames, it has had a hard time finding success in computer wargames. The company's been at it since 1980, and persistence should eventually pay off.

**CompuServe.** Similar to GEnie (see below).

**Broderbund.** Its games are striking visually, but not many wargames.

**Dynamix.** One of the major players in the wargame simulation-game genre.

**Electronic Arts (EA).** Something of a giant in the industry, mainly because it functions largely as a distributor for small "production companies" (that just develop the software). It has numerous subsidiary labels. Because there is no centralized control over product development and quality, EA's games range from the sublime to the abysmal.

**General Quarters.** Noted for its naval wargames.

**GEnie.** An on-line computer service, accessed with a PC and modem, which offers a number of wargames allowing hundreds of gamers to play simultaneously against each other. A much larger selection of on-line games than the other computer services. It also provides bulletin boards and electronic mail that provide an increasingly popular way to play games and exchange information.

**Interstel.** One hit (*Empire*), and a lot of near misses.

**Koei.** Publishes English versions of Japanese computer wargames.

**Kesmai.** Develops wargames for on-line services (GEnie and CompuServe).

**LucasFilm.** Specializes in aircraft simulators and is one of the leaders in this field.

**MicroProse.** Another outfit that has done well with wargame simulators (air, land, and naval). Pioneered the use of extensive printed historical commentary. It has set a standard many other publishers are aiming for. A number of paper-wargame developers ended up here, and it shows.

**Mindscape.** A few attempts at the wargame genre.

**Quantum Quality Productions (QQP).** A new company, which started out by publishing two wargames in 1991 that stressed "game" over "war" and got away with it. Nice work and bodes well for the future.

**RAW.** Like EA, mainly a distribution company. Has obtained the rights to several detailed computer wargames that were unable to obtain wide distribution on their own. RAW had the clout to get these games into a lot of stores. Nice concept.

**Simulations Canada.** An odd duck if there ever was one. Began in the 1970s as a paper-wargame publisher. Located in Nova Scotia, mainly because Canadian government provided financial incentives for new companies that did so. Switched to computer wargames in the 1980s, but its games were not as flashy as the usual computer wargame, and many still depend on paper play aids (as in a paper map).

**Spectrum Holobyte.** Another of the wargame-simulator outfits. Still in the running, although it rarely comes in first.

**Spinnaker.** A wargame once in a while.

**Strategic Simulations, Inc.** One of the earliest computer-wargame publishers that was noted for its direct takeoffs of existing paper wargames. For that reason, it was often called "Stolen Simulations, Incorporated." The only problem with that concept (which was not illegal, just obvious) was that it did not use the PC to greatest effect. SSI eventually became the official publisher of *Dungeons and Dragons* computerized role-playing games. This kept SSI in business and enabled it to continue to publish a large number of excellent computer wargames. Mostly original work these days.

**Strategic Studies Group.** An Australian computer-wargame publisher that stresses the use of standard game systems that, for different battles, are modified to fit each unique situation. Very prolific publisher of nonsimulator computer wargames.

**SubLogic.** The original aircraft-simulator publisher. There's always a chance for a comeback as it's still in business.

**Taito.** A Japanese computer-wargame company, publishing English translations of computer wargames originally published in Japan.

**Three Sixty Pacific (360).** Made it big in 1988 with computer version of GDW naval miniature rules (*Harpoon*). Has had mixed success since then, but is still turning out interesting products that sell.

For all types of wargames, there are many other publishers that have not been mentioned if only because many of the smaller ones come and go so quickly. The best opportunity to see some of the more active (or at least currently active) publishers is to go to a software or game store and look around. Go to one of the wargame conventions, and you can usually meet the designers and publishers in person. Also, many of the magazines mentioned carry reviews, and ads for direct-mail vendors of all types of wargames.

# ✱ COMPUTER WARGAMES PUBLISHED: 1979–1991 ✱

Computer wargames published to date, in alphabetical order. The following list is not complete, partially because I left out some games that were arguable wargames and because I also left out some real dogs that either had limited distribution (or should have had limited distribution). The list is primarily limited because I had to make quite a pitch to my editor and the marketing people to get as much space as you find here. Since 1978, over seven hundred computer wargames have been published. Most of them are listed here, and all the important (popular or just plain well done) ones are listed. I may be off on some of the publication dates. There may also be a few cases where I have the wrong publication dates. There may also be a few cases where I have the wrong publisher listed. The ''author'' is another problem, as computer games tend to be created by teams. Wherever possible, I listed either the sole author (designer and programmer) or the lead person of a team. Well, it's a start. If it's any help, you may encounter a public domain version of this list floating around BBSs or places like GEnie. I obtained a lot of info by doing that for over a year and I will continue to update and support the public domain version.

| Title | Publisher | Author | Date |
|---|---|---|---|
| 688 Attack Sub | Electronic Arts | Ratcliff | 1990 |
| A T A C | Microplay | Internal | 1992 |
| A-10 Tank Killer | Dynamix | Slye | 1990 |
| Abrams Battle Tank I | Electronic Arts | Slye | 1989 |
| Ace Of Aces | Accolade | Bates | 1986 |
| Ace: Air Combat Emulator | Spinnaker | Internal | 1987 |
| Aces Of The Pacific | Dynamix | Slye | 1992 |
| Aces Over Europe | Sierra | Internal | 1993 |
| Action In The North Atlantic | GQ | Hall | 1989 |
| Action Off The River Plate | GQ | Hall | 1987 |
| Action Stations I | RAW | Zimm | 1989 |
| Afrika Korps | Impress | Wright | 1991 |
| After Pearl | Superware | Schwenk | 1984 |
| Air Duel | Microprose | Williams | 1989 |
| Air Force Command | Impressions | Internal | 1992 |
| Air Raid Pearl Harbor | GQ | Hall | 1991 |
| Air Strike USA | Cinemaware | Swift | 1990 |
| Air Traffic Control Simulator | HJC | Internal | 1988 |
| Air Warrior | GEnie | Kesmai | 1988 |
| Airborne Ranger | Microprose | Schick | 1987 |
| Airsim-3 | M Systems | Internal | 1985 |
| American Civil War Vol 1 | SSG | Keating | 1988 |
| American Civil War Vol 2 | SSG | Keating | 1989 |
| American Civil War Vol 3 | SSG | Keating | 1989 |
| Ancient Art Of War At Sea | Broderbund | Murray | 1987 |
| Ancient Art Of War In The Skies | Broderbund | Murray | 1992 |
| Ancient Art of War | Broderbund | Internal | 1986 |
| Ancient Battles | CCS | Smith | 1989 |
| Annals Of Rome | PSS | Internal | 1988 |
| Antietam | SSI | Kroegel | 1985 |
| Apache Strike | Electronic Arts | Sconbeach | 1988 |
| Armada | PSS | Internal | 1990 |
| Armada 2525 | Interstell | Internal | 1991 |
| Armor Alley | 360 | Britto | 1990 |
| Armor Assault | Epyx | Weber | 1983 |
| Arnhem - The Market-Garden Operation | CSS | Smith | 1991 |
| At The Gates Of Moscow | SGP | Heath | 1985 |
| Austerlitz | Cornerstone | Beckett | 1990 |
| Avenger A-10 I | Spectrum Holobyte | Bayless | 1992 |
| B-1 Nuclear Bomber | Avalon Hill | Internal | 1980 |
| B-17 Flying Fortress | Microprose | Internal | 1992 |

| Title | Publisher | Author | Date |
|---|---|---|---|
| B-24 Combat Simulator | SSI | Gray | 1987 |
| Balance Of Power | Mindscape | Chrawford | 1985 |
| Balance Of The Planet | Accolade | Crawford | 1990 |
| Balance of Power: 1990 | Mindscape | Crawford | 1988 |
| Baltic 1985 | SSI | Internal | 1984 |
| Bandit Kings Of Ancient China I | Koei | Shibusawa | 1989 |
| Banzai | GQ | Hall | 1991 |
| Battalion Commander | SSI | Hille | 1983 |
| Battle Command | TTR | Internal | 1986 |
| Battle Cruiser | SSI | Grigsby | 1987 |
| Battle For Midway | Firebird | Internal | 1987 |
| Battle Front | SSG | Keating | 1987 |
| Battle Group | SSI | Grigsby | 1986 |
| Battle Hawks 1942 | Lucasfilm | Falstein | 1988 |
| Battle Isle | UBI | Internal | 1991 |
| Battle Of Britain | Firebird | Internal | 1990 |
| Battle Of The Atlantic | Simcan | Howie | 1986 |
| Battle Of The Bulge | Ark Royal | Carpenter | 1988 |
| Battle Of Waterloo | SSI | Internal | 1980 |
| Battle Stations | GQ | Hall | 1991 |
| Battle Tank: Barbarossa-Stalingrad | Simcan | Newburg | 1990 |
| Battle Tank: Kursk-Berlin | Simcan | Newburg | 1991 |
| Battlefield 2000 | Nova | Internal | 1993 |
| Battlefront | SSG | Keating | 1987 |
| Battleground I | MVP | Internal | 1987 |
| Battlehawks 1942 | Lucasfilm | Internal | 1991 |
| Battles Of Napoleon | SSI | Kroegel | 1988 |
| Battleship Bismarck | GQ | Hall | 1991 |
| Berlin 1948 | Electric Zoo | Internal | 1990 |
| Beyond Castle Wolfenstein | Muse | Internal | 1984 |
| Big Three | SDJ | Internal | 1992 |
| Birds Of Prey | Electronic Arts | Internal | 1991 |
| Bismarck: The North Sea Chase | Datasoft | Stoddart | 1987 |
| Blitzkrieg At The Ardennes | RAW | Benincasa | 1989 |
| Blitzkrieg May 1940 | Impressions | Internal | 1988 |
| Blue Angels | Accolade | Banks | 1990 |
| Blue Max | Synapse | Internal | 1983 |
| Blue Max | 360 | Banks | 1990 |
| Blue Powder Gray Smoke | Garde | Bosson | 1987 |
| Bomb Alley | SSI | Grigsby | 1982 |
| Bomber | Activision | Vektor | 1986 |

| Title | Publisher | Author | Date |
|---|---|---|---|
| Borodino | PSS | Internal | 1990 |
| Borodino: 1812 | Krentek | Beckett | 1987 |
| Bravo Romeo Delta | Ind. | Internal | 1992 |
| Breach 1 | Omnitrend | Leslie | 1987 |
| Breach 2 | Omnitrend | Internal | 1991 |
| Breakthrough In The Ardennes | SSI | Landrey | 1983 |
| Brigade Commander | TTR | Rush | 1991 |
| Broadsides | SSI | Internal | 1983 |
| By Fire & Sword | Avalon Hill | Internal | 1985 |
| Campaign | Readysoft | Internal | 1993 |
| Cardinal Of The Kremlin | Capstone | Internal | 1991 |
| Carrier Command | Microplay | Edgeley | 1988 |
| Carrier Force | SSI | Grigsby | 1983 |
| Carrier Force | SSI | Grigsby | 1992 |
| Carriers At War | SSG | Keating | 1984 |
| Carriers At War | SSG | Keating | 1991 |
| Carthage | Pygnosis | Internal | 1991 |
| Castle Wolfenstein | Muse | Internal | 1983 |
| Castles | Interplay | Fargo | 1991 |
| Castles II | Interplay | Fargo | 1993 |
| Centurion: Defender Of Rome | Electronic Arts | Beck | 1990 |
| Charge Of The Light Brigade, The | Impressions | Internal | 1990 |
| Chickamauga | GDW | Internal | 1985 |
| Chickamauga, Rebel Charge At | SSI | Kroegel | 1987 |
| Civil War I | Avalon Hill | Estvanik | 1988 |
| Civilization | Microprose | Meier | 1991 |
| Clash Of Wills | DKG | Summerlott | 1985 |
| Clear For Action | Avalon Hill | Internal | 1984 |
| Clipper | PBI | Internal | 1982 |
| Close Assault | Avalon Hill | Bedrosian | 1983 |
| Cohorts | Impressions | Internal | 1989 |
| Colonial Conquest | SSI | Cermak | 1985 |
| Combat Leader | SSI | Hille | 1983 |
| Command HQ | Microprose | Bunten | 1990 |
| Computer Air Combat | SSI | Merrow | 1980 |
| Computer Ambush | SSI | Williger | 1980 |
| Computer Bismarck | SSI | Billings | 1980 |
| Computer Conflict | SSI | Keating | 1981 |
| Conflict Europe | Mirrorsoft | Internal | 1988 |
| Conflict In Vietnam | Microprose | Meier | 1986 |
| Conflict: Europe | Mirrorsoft | Steel | 1989 |

| Title | Publisher | Author | Date |
|---|---|---|---|
| Conflict: Korea | SSI | Internal | 1992 |
| Conflict: Middle East | SSI | Koger | 1991 |
| Conquest of Japan | Impressions | Internal | 1992 |
| Conquests of the Longbow | Sierra | Internal | 1991 |
| Covert Action | Microprose | Meier | 1990 |
| Crisis In The Kremlin | Spectrum Holobyte | Internal | 1991 |
| Crusade In Europe | Microprose | Meier | 1985 |
| D-Day | Ark Royal | Internal | 1987 |
| Dam Busters | Accolade | Sydney | 1985 |
| Das Boot | 360 | Butler | 1991 |
| Dawn Patrol | TSR | Internal | 1983 |
| Decision At Gettysburg | Tiglon | Internal | 1990 |
| Decision In The Desert | Microprose | Meier | 1985 |
| Decisive Battles Of The Civil War | SSG | Keating | 1988 |
| Defcon 5 | Cosmi | Norman | 1988 |
| Defender Of The Crown | Cinemaware | Beeck | 1987 |
| Desert Fox | Accolade | Sydney | 1987 |
| Desert Rats | CCS | Smith | 1989 |
| Desert War | SJP | Internal | 1989 |
| Destroyer | Epyx | Internal | 1986 |
| Destroyer Escort | Medalist | Internal | 1989 |
| Diplomacy | Avalon Hill | Internal | 1984 |
| Diplomacy (92 edition) | Avalon Hill | Internal | 1992 |
| Discovery | Impressions | Internal | 1992 |
| Dive Bomber | US Gold | Internal | 1989 |
| Dnieper River Line | Avalon Hill | Ketchledge | 1983 |
| Dragon Force | Interstel | Damon | 1990 |
| Dreadnoughts | Avalon Hill | Dowell | 1984 |
| Dreadnoughts | Impressions | Internal | 1992 |
| Eagles | SSI | Raymond | 1983 |
| Eagles Nest, Into The | Mindscape | Internal | 1987 |
| Eastern Front | Apx | Crawford | 1982 |
| Elite Plus | Microprose | Internal | 1988 |
| Empire | Interstel | Baldwin | 1987 |
| Europe Ablaze | SSG | Keating | 1985 |
| F-117A | Microprose | Briggs | 1991 |
| F-14 Tomcat | Activision | Dynamix | 1989 |
| F-15 Strike Eagle | Microprose | Meier | 1987 |
| F-15 Strike Eagle 2 | Microprose | Internal | 1991 |
| F-15 Strike Eagle 3 | Microprose | Meier | 1992 |
| F-16 Combat Pilot | Electronic Arts | Digital In | 1990 |

| Title | Publisher | Author | Date |
|---|---|---|---|
| F-18 Interceptor | Electronic Arts | Dinnerman | 1989 |
| F-19 | Microprose | Meier | 1988 |
| F-29 | Ocean | Kenwright | 1990 |
| Falcon | Spectrum Holobyte | Louie | 1987 |
| Falcon 3.0 | Spectrum Holobyte | Louie | 1991 |
| Falcon AT | Spectrum Holobyte | Louie | 1989 |
| Falklands '82 | Firebird | Internal | 1988 |
| Fall Gelb | Simcan | Internal | 1988 |
| Feudal Lords | Impressions | Internal | 1991 |
| Field Of Fire | SSI | Damon | 1984 |
| Field of Fire | SSI | Internal | 1986 |
| Fifth Eskrada | Simcan | Nichols | 1990 |
| Fifty Mission Crush | SSI | Gray | 1984 |
| Fighter Command | SSI | Merrow | 1983 |
| Fighter Duel | Jaeger | Internal | 1992 |
| Fighting Men of Rome | Merit | Internal | 1991 |
| Final Conflict | Merit | Internal | 1992 |
| Fire Brigade | Panther | O'Connor | 1989 |
| Firepower | Microillu. | Wolfshield | 1987 |
| Firepower | Activision | Wolfshield | 1990 |
| Fireteam 2200 | RAW | Internal | 1989 |
| First Over Germany | SSI | Gray | 1989 |
| Fleet Commander | RAW | Internal | 1993 |
| Fleet Med | Simcan | Internal | 1991 |
| Flight Of The Intruder | Spectrum Holobyte | Hyde | 1990 |
| Flight Simulator | Sublogic | Artwick | 1979 |
| Flight Simulator | Microsoft | Artwick | 1983 |
| Flight Simulator 4.0 | Microsoft | Internal | 1991 |
| Flight of the Intruder | Spectrum Holobyte | Internal | 1991 |
| Flying Tigers | Discovery | Weseley | 1982 |
| Fokker Triplane Simulator | Bullseye | Internal | 1986 |
| Fokker Triplane Simulator | Bullseye | Internal | 1989 |
| Fort Apache | Impressions | Internal | 1991 |
| Frontline | CCS | Lenton | 1990 |
| Future Wars | Interplay | Internal | 1990 |
| Galleons Of Glory | Broderbund | Portwood | 1991 |
| Gato | Spectrum Holobyte | Dawson | 1983 |
| Genghis Khan | Koei | Internal | 1989 |
| Geopolitique 1990 | SSI | Ketchledge | 1983 |
| German Raider Atlantis | GQ | Hall | 1989 |
| Germany 1985 | SSI | Keating | 1983 |

| Title | Publisher | Author | Date |
|---|---|---|---|
| Gettysburg | SSI | Kroegel | 1986 |
| Gettysburg:The Turning Point | SSI | Internal | 1989 |
| Global Commander | Datasoft | Wilson | 1988 |
| Global Confrontation | Aigis Cir | Internal | 1987 |
| Global Conquest | Miller | Internal | 1992 |
| Global Effect | Impressions | Internal | 1992 |
| Golan Front | Simcan | Nichols | 1985 |
| Gold Of The Americas | SSG | Hart | 1989 |
| Gorbachev's Ace | Electronic Arts | Internal | 1987 |
| Grand Fleet | Simcan | Baker | 1988 |
| Gray Storm Rising | Ark Royal | Internal | 1989 |
| Great Battles | Royal | Internal | 1990 |
| Great Naval Battles | SSI | Internal | 1993 |
| Great War 1914 | DKG | Summerlott | 1986 |
| Grey Seas Grey Skies | Simcan | Internal | 1989 |
| Guadalcanal Campaign | SSI | Grigsby | 1982 |
| Guderian | Avalon Hill | SW Asso. | 1987 |
| Gulf Strike | Avalon Hill | Herman | 1987 |
| Gunboat | Accolade | Loughry | 1990 |
| Guns & Butter | Mindscape | Crawford | 1990 |
| Gunship | Microprose | Hendrick | 1986 |
| Gunship 2000 | Microprose | Day | 1991 |
| Halls Of Montezuma | SSG | Keating | 1988 |
| Harpoon | 360 | Walton | 1988 |
| Harpoon II | 360 | Internal | 1993 |
| Harrier | Accolade | Internal | 1989 |
| Harrier Combat Simulator | Mindscape | Internal | 1988 |
| Heart Of Africa | Electronic Arts | Internal | 1985 |
| Heavy Metal | Access | Erickson | 1989 |
| Hellcat Ace | Microprose | Meier | 1983 |
| Heroes of the 357th | Electronic Arts | Internal | 1992 |
| Hidden Agenda | Springboard | Gasperini | 1989 |
| High Seas | Garde | Bosson | 1987 |
| Hundred Years War | GEnie | Dunnigan | 1992 |
| Hunt For Red October | Datasoft | Oxford | 1990 |
| Hurricane | Simcam | Internal | 1990 |
| In Harms Way | Simcan | Nichols | 1988 |
| Incunabula | Avalon Hill | Internal | 1985 |
| Iwo Jima/Falklands | Firebird | Bethell | 1985 |
| Jagdstaffel | Discovery | Weseley | 1985 |
| Jet | Sublogic | Guy | 1987 |

| Title | Publisher | Author | Date |
|---|---|---|---|
| Jet Combat Simulator | Epyx | Internal | 1989 |
| Jet Fighter 2.0 | SubLogic | Internal | 1992 |
| Jetfighter I | Broderbund | Dinnerman | 1988 |
| Jetfighter II | Velocity | Internal | 1991 |
| Joan Of Arc: Siege & The Sword | Broderbund | Perconti | 1989 |
| Jumpjet | Eurosoft | Internal | 1990 |
| Kampfgruppe | SSI | Grigsby | 1985 |
| Knights Of The Desert | SSI | TDG | 1983 |
| Knights Of The Sky | Microprose | Briggs | 1991 |
| Kriegsmarine | Simcan | Newberg | 1989 |
| Kursk Campaign | Simcan | Internal | 1986 |
| L'Empereur | Koei | Internal | 1991 |
| LHX | Electronic Arts | Internal | 1990 |
| Lafayette Escadrille | Discovery | Weseley | 1983 |
| Legionnaire | Avalon Hill | Crawford | 1988 |
| Legions Of Rome | Circle | Internal | 1990 |
| Liberty or Death | SSI | Bever | 1990 |
| Line in the Sand | SSI | Internal | 1992 |
| Long Lance | Simcan | Nichols | 1988 |
| Lords Of Conquest | Electronic Arts | Internal | 1986 |
| Lords Of The Rising | Cinemaware | Barnett | 1989 |
| Lost Admiral | QQP | Zaccagnin | 1991 |
| Lost Patrol | Ocean | Harling | 1989 |
| Luftflotte | Ark Royal | Internal | 1989 |
| M-1 Tank Platoon | Microprose | Hendrick | 1989 |
| M.U.L.E | Electronic Arts | Bunten | 1986 |
| Macarthur's War | SSG | Keating | 1989 |
| Macarthur's War | SSG | Keating | 1991 |
| Mach 2 Combat Flight Simulator | Firebird | Internal | 1990 |
| Main Battle Tank: Central Germany | Simcan | Newberg | 1989 |
| Main Battle Tank: Middle East | Simcan | Internal | 1992 |
| Main Battle Tank: Northern Germany | Simcan | Internal | 1990 |
| Malta Storm | Simcan | Crandall | 1990 |
| Malta Strike | Discovery | Arneson | 1980 |
| Mantis | Oaragon | Internal | 1993 |
| Mare Nostrum | GQ | Hall | 1990 |
| Marianas Turkey Shoot | GQ | Hall | 1990 |
| Maximum Overkill | Nova | Internal | 1993 |
| Mech Brigade | SSI | Grigsby | 1985 |
| Mechwarrior | Activision | Internal | 1990 |
| Medieval Lords | SSI | Campion | 1991 |

| Title | Publisher | Author | Date |
|---|---|---|---|
| Medieval Warriors | Merit | Internal | 1991 |
| Megafortress | 360 | Internal | 1991 |
| Merchant Colony | Impressions | Internal | 1991 |
| Metz-Cobra | KG | Internal | 1989 |
| MidWinter | Microplay | Singleton | 1990 |
| Midway Campaign | NMA | Internal | 1980 |
| Mig Alley | Microprose | Hollis | 1987 |
| Mig-29 Fulcrum | Domark | Kavanagh | 1991 |
| Migs And Messerschmitts | Discovery | Maker | 1980 |
| Miracle At Midway | GQ | Hall | 1989 |
| Modem Wars | Electronic Arts | Bunten | 1989 |
| Moscow Campaign | Simcan | Internal | 1987 |
| NATO Commander | Microprose | Internal | 1985 |
| Nam | SSI | Internal | 1986 |
| Napoleon At Waterloo | Krentek | Internal | 1985 |
| Napoleon's Campaigns | SSI | Internal | 1981 |
| Navy SEAL | Cosmi | Internal | 1989 |
| No Greater Glory | SSI | Bever | 1991 |
| Nobunaga's Ambition | Koei | Swartz | 1988 |
| Nobunaga's Ambition 2 | Koei | Internal | 1991 |
| Normandy Battle For | SSI | Landrey | 1984 |
| Normandy Battles In | SSG | Keating | 1987 |
| North & South | DataEast | Baudet | 1990 |
| North Atlantic | SSI | Grigsby | 1983 |
| North Atlantic Convoy Raider | Avalon Hill | Internal | 1987 |
| Northern Fleet | Simcan | Internal | 1989 |
| Norway '85 | SSI | Keating | 1985 |
| Nuclear War | New World | Hyman | 1990 |
| Nukewar | Avalon Hill | Internal | 1980 |
| Objective: Kursk | SSI | Grigsby | 1984 |
| Ocean Ranger | Activision | Patrick | 1988 |
| Ogre | Origin | Internal | 1986 |
| Old Ironsides | Xerox | Rice | 1988 |
| Omega | Origin | Internal | 1989 |
| Operation Apocalypse | SSI | Internal | 1989 |
| Operation Combat | Merit | Internal | 1990 |
| Operation Keystone | SSI | Peto | 1986 |
| Operation Market-Garden | SSI | Kroegel | 1986 |
| Operation Overlord | Simcan | Internal | 1986 |
| Operation Whirlwind | Broderbund | Damon | 1984 |
| Operation: Combat | Merit | Lamb | 1990 |

| Title | Publisher | Author | Date |
|---|---|---|---|
| Orbiter | Spectrum Holobyte | Internal | 1985 |
| Overlord | CSS | Wright | 1989 |
| Overrun | SSI | Grigsby | 1989 |
| P-51 Mustang | Bullseye | Internal | 1989 |
| PHM Pegasus | Lucasfilm | Falstien | 1986 |
| PT-109 | Spectrum Holobyte | Walton | 1988 |
| Pacific Island | Empire | Internal | 1992 |
| Pacific Storm: The Solomons | Simcan | Internal | 1991 |
| Pacific Storm: Midway | Simcan | Internal | 1991 |
| Pacific War | Koei | Internal | 1992 |
| Panzer Battles | SSG | Keating | 1989 |
| Panzer Grenadier | SSI | Damon | 1985 |
| Panzer Strike | SSI | Grigsby | 1988 |
| Panzer War | Windcrest | Internal | 1984 |
| Panzerjagd | Avalon Hill | Scorupski | 1984 |
| Paris In Danger | Avalon Hill | Internal | 1989 |
| Parthian Kings | Avalon Hill | Internal | 1987 |
| Patriot | 360 | Internal | 1992 |
| Patton Strikes Back | Broderbund | Crawford | 1991 |
| Patton Vs Rommel | Mindscape | Crawford | 1987 |
| Pegasus | Lucasfilm | Falstein | 1986 |
| Perfect General | QQP | Baldwin | 1991 |
| Pirates | Microprose | Meier | 1987 |
| Platoon | DataEast | Internal | 1989 |
| Populous | Bullfrog | Internal | 1990 |
| Power At Sea | Accolade | Internal | 1987 |
| Powermonger | Bullfrog | Internal | 1992 |
| Prelude To Jutland | GQ | Hall | 1986 |
| Project Stealth Fighter | Microprose | Synoski | 1987 |
| Pursuit Of The Graf Spee | SSI | Billings | 1982 |
| RAF in the Pacific | Sierra | Internal | 1993 |
| RAF: Battle Of Britain | Discovery | Arneson | 1979 |
| RAM | Avalon Hill | Estvanik | 1985 |
| RDF | SSI | Keating | 1990 |
| Railroad Tycoon | Microprose | Meier | 1990 |
| Rails West | SSI | Internal | 1984 |
| Reach For The Skies | Rowan | Internal | 1989 |
| Red Baron | Dynamix | Slye | 1991 |
| Red Lightning | SSI | Koger | 1989 |
| Red Phoenix | Mirrorsoft | Internal | 1992 |
| Red Storm Rising | Microprose | Meier | 1989 |

| Title | Publisher | Author | Date |
|---|---|---|---|
| Reforger 88 | SSI | Grigsby | 1984 |
| Return To The Falklands | GQ | Hall | 1987 |
| Revolution '76 | Britannica | Bever | 1989 |
| Rising Sun The | GQ | Hall | 1989 |
| Road To Gettysburg | SSI | Murray | 1982 |
| Road To Moscow | GDW | Gardocki | 1987 |
| Rock Of Stalingrad The | Benchmark | Newburg | 1983 |
| Romance Of The Three Kingdoms 1 | Koei | Swartz | 1988 |
| Romance Of The Three Kingdoms 2 | Koei | Internal | 1991 |
| Rome And The Barbarians | Krentek | Internal | 1984 |
| Rommel At El Alamein | Simcan | Nichols | 1988 |
| Rommel At Gazala | Simcan | Internal | 1986 |
| Rommel Battles For N Africa | SSG | Keating | 1988 |
| Rommel In The Battlefront | Infra | Internal | 1987 |
| Rommel at El Alamein | SimCan | Nichols | 1988 |
| Rommel: Battles For North Africa | SSG | Keating | 1988 |
| Rommel: Battles For Tobruk | GDW | Chadwick | 1986 |
| Rourke's Drift | Impressions | Internal | 1989 |
| Rules of Engagement | Omnitrend | Internal | 1988 |
| Russia | SSG | Keating | 1987 |
| SDI | Cinemaware | Beeck | 1987 |
| Sands Of Fire | 360 | Walton | 1990 |
| Saratoga | Apx | Internal | 1988 |
| Second Front | SSI | Grigsby | 1990 |
| Secret Weapons Of The Luftwaffe | Lucasfilm | Holland | 1991 |
| Seven Cities Of Gold | Electronic Arts | Internal | 1984 |
| Seventh Fleet | Simcan | Nichols | 1985 |
| Sherman M4 | Lorelei | Arnaud | 1991 |
| Shiloh The Battle Of | SSI | Landrey | 1981 |
| Shiloh: Grant's | SSI | Landrey | 1987 |
| Sieg In Afrika | Simcan | Internal | 1984 |
| Siege | Mindcraft | Internal | 1992 |
| Silent Service | Microprose | Meier | 1985 |
| Silent Service 2 | Microprose | Hendrick | 1990 |
| Simearth | Maxis | Wright | 1991 |
| Six Gun Shootout | SSI | Internal | 1985 |
| Sky Chase | Maxis | Internal | 1989 |
| Sky Shark | Taito | Internal | 1989 |
| Sniper | Compuserve | Estvanik | 1988 |
| Sonar Search | CC | Internal | 1984 |
| Sons of Liberty | SSI | Landrey | 1987 |

| Title | Publisher | Author | Date |
|---|---|---|---|
| Sorcerer Lord | PSS | Internal | 1988 |
| Southern Command | SSI | Keating | 1982 |
| Spitfire | Avalon Hill | Virgin | 1986 |
| Spitfire Ace | Microprose | Meier | 1984 |
| Spitfire Simulator | Mind Syste | Internal | 1988 |
| Spoils of War | RAW | Internal | 1991 |
| Stalingrad Campaign | Simcan | Nichols | 1987 |
| Starfleet 1 | Interstel | Internal | 1987 |
| Starfleet 2 | Interstel | Internal | 1990 |
| Stealth Mission | Sublogic | Setzler | 1988 |
| Steel Thunder | Accolade | Loughry | 1988 |
| Storm Across Europe | SSI | Cermak | 1989 |
| Stormovik | Electronic Arts | Grace | 1990 |
| Strike Aces | Accolade | Craven | 1990 |
| Strike Commander | Origin | Internal | 1992 |
| Strike Fleet I | Lucasfilm | Falstein | 1987 |
| Strike Force Harrier | Escape | Internal | 1989 |
| Sub Battle Simulator | Epyx | Walton | 1987 |
| Suez 73 | RAW | Benincasa | 1991 |
| Super Huey | Cosmi | Norman | 1989 |
| Surrender At Stalingrad | DKG | Summer | 1986 |
| Sword Of Aragon | SSI | Shilling | 1989 |
| Sword Of The Samurai | Microprose | Schick | 1989 |
| TAC | Avalon Hill | Bosson | 1984 |
| Tank | Spectrum Holobyte | Widjaja | 1989 |
| Tank Platoon | Dataworks | Internal | 1990 |
| Tanks And Squads | C&C | Winfree | 1982 |
| Tanktics | Avalon Hill | Crawford | 1981 |
| Task Force 1942 | Microprose | Internal | 1992 |
| Taskforce | Jagdstaffel | Internal | 1983 |
| Team Yankee | Empire | Green | 1990 |
| The Great Escape | Thunder | Internal | 1986 |
| The Train: Escape To Normandy | Accolade | Artech | 1988 |
| Theater Europe | Datasoft | Steels | 1985 |
| Theater of War | 360 | Internal | 1992 |
| Their Finest Hour | Lucasfilm | Holland | 1989 |
| Third Reich | Avalon Hill | Thalean | 1992 |
| Thud Ridge | 360 | Internal | 1988 |
| Tigers In The Snow | SSI | Kroegel | 1981 |
| To The Rhine | Simcan | Newburg | 1987 |
| Tobruk: The Clash Of Armour | Datasoft | Williams | 1987 |

| Title | Publisher | Author | Date |
|---|---|---|---|
| Tomahawk | Datasoft | Marshall | 1990 |
| Torpedo Fire | SSI | Lyon | 1981 |
| Tracon | Wesson | Wesson | 1989 |
| Tracon 2 | Wesson | Internal | 1992 |
| Tsushima | Avalon Hill | Kiya | 1985 |
| Twilight 2000 | Paragon | Chadwick | 1991 |
| Twilight 2001 | Microplay | Chadwick | 1992 |
| Typhoon of Steel | SSI | Grigsby | 1988 |
| UMS | Firebird | Sidran | 1987 |
| UMS 2 | Microplay | Sidran | 1991 |
| USAAF | SSI | Grigsby | 1985 |
| Under Fire | Avalon Hill | Bosson | 1985 |
| Under Southern Skies | Avalon Hill | Hall | 1984 |
| Under The Ice | Lyric | Almberg | 1989 |
| Up Periscope | Actionsoft | Internal | 1986 |
| V for Victory: Utah Beach | 360 | Internal | 1991 |
| V for Victory: Velikiye Luki | 360 | Internal | 1992 |
| VC | Avalon Hill | Monk | 1989 |
| Victory at Sea | 360 | Dunnigan | 1993 |
| Vikings | 21st Century | Internal | 1992 |
| War At Sea | GQ | Hall | 1988 |
| War In Russia | SSI | Grigsby | 1984 |
| War In The Falklands | GQ | Hall | 1986 |
| War In The Pacific | SSI | Grigsby | 1992 |
| War In The South Pacific | SSI | Grigsby | 1987 |
| Wargame Construction Set | SSI | Damon | 1986 |
| Warlords | SSG | Fawkner | 1991 |
| Warship | SSI | Grigsby | 1986 |
| Warship That Changed History, The | GQ | Hall | 1987 |
| Warthog A-10 | Spectrum Holobyte | Internal | 1993 |
| Waterloo | SSI | Turcan | 1990 |
| Western Front | SSI | Grigsby | 1991 |
| White Death | RAW | Benincasa | 1988 |
| Wing Commander | Origin | Internal | 1989 |
| Wing Commander 2 | Origin | Internal | 1991 |
| Winged Samurai | Discovery | Maker | 1980 |
| Wingman | Microprose | Internal | 1988 |
| Wings | Cinemaware | Cutter | 1990 |
| Wings Of Fury | Broderbund | Waldo | 1987 |
| Wings Of War | SSI | Merrow | 1985 |
| Wolfpack | Broderbund | Garcia | 1990 |

| Title | Publisher | Author | Date |
|---|---|---|---|
| Wolfstein 3-D | Apogee | Internal | 1992 |
| Wooden Ships & Iron Men | Avalon Hill | Taylor | 1986 |
| World War I | Wipeout | Internal | 1988 |
| World War II:1946 | Sierra | Internal | 1993 |
| Worlds At War | RAW | Internal | 1991 |
| Yeager AFT 2.0 | Electronic Arts | Internal | 1991 |
| Yeager's Advanced Flight Trainer | Electronic Arts | Learner | 1990 |
| Yeager's Air Combat | Electronic Arts | Iverson | 1991 |

## ✱ ✱ QUESTIONAIRE ✱ ✱

There's a good chance that there will be a third edition of this book in the future, so I thought it prudent to get some feedback from users of this edition. To participate, fill out the following questionnaire (either the original or a photocopy) and mail to me at 328 West 19th Street, NY, NY 10011. Results will be published and discussed in the next edition. Perhaps next century; see you then. Rate each of the chapters on a scale of 1 to 9 (1 = liked very little, 9 = liked very much).

**1.** _______ **1.** What Is a Wargame?

**2.** _______ **2.** How to Play

**3.** _______ **3.** Why Play the Games

**4.** _______ **4.** Designing Wargames

**5.** _______ **5.** History of Wargames

**6.** _______ **6.** Computer Wargames

**7.** _______ **7.** Designing Computer Wargames

**8.** _______ **8.** Who Plays the Games

**9.** _______ **9.** Wargames at War

**10.** _______ Appendix: Reading List

**11.** _______ Appendix: My Game Bibliography

**12.** _______ Appendix: Gaming Aids

**13.** _______ Appendix: Computer Wargames Published: 1979–1991

**14.** _______ Appendix: Paper Wargame Publishers

**15.** _______ Appendix: Computer Wargame Publishers

**Write in any suggestions for the next edition.**

# ✷ ✷ INDEX ✷ ✷